my **revisi⊙n** notes

OCR AS/A-level Year 1

BIOLOGY A

Frank Sochacki

HODDER
EDUCATION
AN HACHETTE UK COMPANY

Hachette UK's policy is to use papers that are natural, renewable and recyclable products and made from wood grown in sustainable forests. The logging and manufacturing processes are expected to conform to the environmental regulations of the country of origin.

Orders: please contact Bookpoint Ltd, 130 Milton Park, Abingdon, Oxon OX14 4SB. Telephone: (44) 01235 827720. Fax: (44) 01235 400454. Email education@bookpoint.co.uk

Lines are open from 9 a.m. to 5 p.m., Monday to Saturday, with a 24-hour message answering service. You can also order through our website: www.hoddereducation.co.uk

ISBN: 978 1 4718 4207 8

© Frank Sochacki 2015

First published in 2015 by

Hodder Education,
An Hachette UK Company
Carmelite House
50 Victoria Embankment
London EC4Y 0DZ

www.hoddereducation.co.uk

Impression number   10 9 8 7 6 5 4 3 2 1

Year   2019 2018 2017 2016 2015

Cover photo reproduced by permission of Kletr/Fotolia

Typeset in Integra Software Services Pvt. Ltd., Pondicherry, India

Printed in Spain

A catalogue record for this title is available from the British Library.

Get the most from this book

Everyone has to decide his or her own revision strategy, but it is essential to review your work, learn it and test your understanding. These Revision Notes will help you to do that in a planned way, topic by topic. Use this book as the cornerstone of your revision and don't hesitate to write in it — personalise your notes and check your progress by ticking off each section as you revise.

Tick to track your progress

Use the revision planner on pages 4 and 5 to plan your revision, topic by topic. Tick each box when you have:

- revised and understood a topic
- tested yourself
- practised the exam questions and gone online to check your answers and complete the quick quizzes

You can also keep track of your revision by ticking off each topic heading in the book. You may find it helpful to add your own notes as you work through each topic.

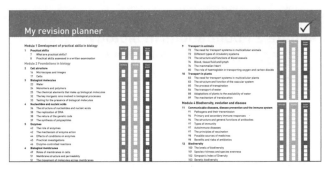

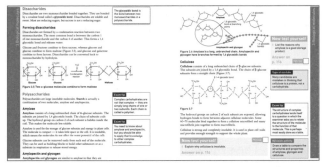

Features to help you succeed

Exam tips

Expert tips are given throughout the book to help you polish your exam technique in order to maximise your chances in the exam.

Typical mistakes

The author identifies the typical mistakes candidates make and explains how you can avoid them.

Now test yourself

These short, knowledge-based questions provide the first step in testing your learning. Answers are at the back of the book.

Definitions and key words

Clear, concise definitions of essential key terms are provided where they first appear.

Key words from the specification are highlighted in bold throughout the book.

Revision activities

These activities will help you to understand each topic in an interactive way.

Exam practice

Practice exam questions are provided for each topic. Use them to consolidate your revision and practise your exam skills.

Summaries

The summaries provide a quick-check bullet list for each topic.

Online

Go online to check your answers to the exam questions and try out the extra quick quizzes at **www.hoddereducation.co.uk/myrevisionnotes**

My revision planner

REVISED TESTED EXAM READY

Exam practice answers and quick quizzes at **www.hoddereducation.co.uk/myrevisionnotes**

Now test yourself answers

**Exam practice answers and quick quizzes at
www.hoddereducation.co.uk/myrevisionnotes**

Countdown to my exams

6–8 weeks to go

- Start by looking at the specification — make sure you know exactly what material you need to revise and the style of the examination. Use the revision planner on pages 4 and 5 to familiarise yourself with the topics.
- Organise your notes, making sure you have covered everything on the specification. The revision planner will help you to group your notes into topics.
- Work out a realistic revision plan that will allow you time for relaxation. Set aside days and times for all the subjects that you need to study, and stick to your timetable.
- Set yourself sensible targets. Break your revision down into focused sessions of around 40 minutes, divided by breaks. These Revision Notes organise the basic facts into short, memorable sections to make revising easier.

REVISED ☐

2–5 weeks to go

- Read through the relevant sections of this book and refer to the exam tips, exam summaries, typical mistakes and key terms. Tick off the topics as you feel confident about them. Highlight those topics you find difficult and look at them again in detail.
- Test your understanding of each topic by working through the 'Now test yourself' questions in the book. Look up the answers at the back of the book.
- Make a note of any problem areas as you revise, and ask your teacher to go over these in class.
- Look at past papers. They are one of the best ways to revise and practise your exam skills. Write or prepare planned answers to the exam practice questions provided in this book. Check your answers online and try out the extra quick quizzes at **www.hoddereducation.co.uk/ myrevisionnotes**
- Use the revision activities to try out different revision methods. For example, you can make notes using mind maps, spider diagrams or flash cards.
- Track your progress using the revision planner and give yourself a reward when you have achieved your target.

REVISED ☐

1 week to go

- Try to fit in at least one more timed practice of an entire past paper and seek feedback from your teacher, comparing your work closely with the mark scheme.
- Check the revision planner to make sure you haven't missed out any topics. Brush up on any areas of difficulty by talking them over with a friend or getting help from your teacher.
- Attend any revision classes put on by your teacher. Remember, he or she is an expert at preparing people for examinations.

REVISED ☐

The day before the examination

- Flick through these Revision Notes for useful reminders, for example the exam tips, exam summaries, typical mistakes and key terms.
- Check the time and place of your examination.
- Make sure you have everything you need — extra pens and pencils, tissues, a watch, bottled water, sweets.
- Allow some time to relax and have an early night to ensure you are fresh and alert for the examination.

REVISED ☐

My exams

AS Biology

Date:..

Time:..

Location:..

A-level Biology

Date:..

Time:..

Location:..

1 Practical skills

What are practical skills?

The development of practical skills is a fundamental part of the study of biology and it should be an integral part of your learning. These skills not only enhance your understanding of the subject, they also serve as a suitable preparation for the demands of studying biology at a higher level.

At AS you will not have a practical endorsement of your practical skills, but the written exams have questions that test your practical skills. These include:

- planning
- implementing
- analysis
- evaluation

The principles of scientific method

REVISED

Scientific method relies upon gathering evidence. This may be qualitative evidence, but is best if it is quantitative. There may be variations in the approach, but all good scientific work follows the same general pattern. Observation and practical experimentation are key to the process.

1 *Observe a phenomenon and ask a question*, for example 'Bananas go soft and brown as they ripen. Why?'

2 *Carry out research to find out what is already known.* Sources include books, websites and scientific journals. You may find others have carried out experiments on the question. Their results are secondary data.

3 *Make a hypothesis based on what you have found out.* Your hypothesis should be clearly stated. You will also need a null hypothesis which can be tested. A null hypothesis is one that you try to disprove through testing.

4 *Use your hypothesis to make a prediction.* For example, your hypothesis may be 'Bananas go soft because starch is converted to sugars.' The null hypothesis would be 'Starch is not converted to sugar as the banana ripens.' Your prediction would be 'As a banana ripens, the concentration of sugars increases and the concentration of starch drops.'

5 *Plan and carry out an experiment to test your prediction.* Your plan must make clear what are the independent and dependent variables and any other possible variables that should be kept constant (controlled). A detailed method should be written that provides clear guidance on all the techniques used, including concentrations and volumes of all solutions used and how other variables are controlled. The method should be clear enough for someone else to copy the experiment. It should also include an assessment of the hazards and risks involved.

6 *Repeat your experiment to see if you can replicate your results.* Repeating is important to see if your data is reliable. It allows you to identify anomalous results and calculate a mean. You may be able to compare your results with those found in your research (the secondary data). Be prepared to modify your original method if your results are not reliable.

7 *Present your raw data in the form of a table.*

8 *Process your data to calculate means or rates.*

9 *Present your data in the form of a graph if possible.* This helps to identify trends.

10 *Analyse your data.* This often requires the use of a suitable statistical test. Tests can be used to determine:
   - whether the data are reliable
   - whether your results are statistically different from the expected results
   - whether there is a correlation between the independent and dependent variables

11 *Draw conclusions.* Do your data support your hypothesis? If your data refute your null hypothesis, this is evidence that your hypothesis is supported. Does your original hypothesis need to be modified? Are other questions posed as a result of this experiment?

12 *Evaluate your method.* Did your method provide reliable data? Were all variables suitably controlled? Did you make any errors in your method? What was the level of precision? Were there any limitations that may have put your data in doubt? What are the uncertainties?

13 *Evaluate your data.* How reliable are your data? Is the level of precision used appropriate? What level of uncertainty is there? Are the data valid?

14 *Peer assessment/review.* Scientists share their work through presentations or publication. Publication may be in a scientific journal or on peer-review websites so that other scientists can read and judge the work. They consider a number of aspects, including whether:
   - the methodology used was sound
   - the data were correctly interpreted
   - conclusions have been properly deduced from the data
   - there is a conflict of interest

Practical skills assessed in a written examination

You will be required to develop a range of practical skills throughout your course in preparation for the written examinations. Clearly, the written examination cannot test your skills in a practical way. There is no practical exam. Your ability to apply your knowledge to practical situations will be tested. This includes questions that test your skills in:
- planning
- implementing
- analysis
- evaluation

Planning

REVISED

1 You may need to solve a problem set in a practical context.
2 You may be asked to make a prediction from a given hypothesis, or you may be given a prediction and asked how you could test it.
3 You should be able to consider the prediction and outline an experiment or investigation that will test the prediction. For example:

   If you are given the prediction that yeast will stop respiring at 45°C, how would you test it?

4 You would need to describe how you would determine when respiration was taking place and how you would determine when it stops. You should then describe a method in which you tested yeast at a range of temperatures around 45°C to determine at what temperature it actually stops respiring.

Exam practice answers and quick quizzes at **www.hoddereducation.co.uk/myrevisionnotes**

Suitable apparatus, equipment and techniques

1 Select apparatus from a range of normal glassware and chemical reagents.
2 As respiration in all organisms releases carbon dioxide, it would be sensible to arrange a test that demonstrates the presence of carbon dioxide such as using lime water or bicarbonate indicator. You would then place a sample of yeast at different temperatures and demonstrate that it produces carbon dioxide at some temperatures but not at others.
3 You need to select apparatus that enables you to change the temperature of the yeast sample and pass the gas produced through the selected reagent.
4 Draw a diagram of the apparatus and how it is set up.

Application of scientific knowledge

1 Your plan should be based on suitable scientific knowledge. This knowledge should be applied to the practical context you are planning.
2 You will need to show that you understand the process of respiration well enough to know that sugars are needed and carbon dioxide is released. The process involves enzymes and it is these that are affected by the change in temperature.

Identification of variables

1 The independent variable is the variable you choose to change during the experiment. Temperature is the independent variable in this example.
2 The dependent variable is the variable that changes as a result of changing the independent variable. Production of carbon dioxide is the dependent variable in this example.
3 Controlled variables are any other factors that might affect the result, such as pH, volumes of solutions used, concentrations of solutions used.

Evaluation of the method

Show that you have considered whether the method is appropriate to test the desired hypothesis. Ask questions about the method.
1 Would your proposed method do the job?
2 Would it provide an answer to the question?
3 What data will you collect?
4 What will you do with the data collected?
5 Are there any limitations?
6 What level of uncertainty is there in the data collected?

In this example, you might ask:
1 Does it detect whether carbon dioxide is given off?
2 Can I alter the temperature or keep it constant?
3 Can the yeast respire if temperature is not limiting?

Your proposed method may not be perfect or provide a full answer. This is fine as long as you recognise this and say so in your evaluation.

Implementing

REVISED

As part of the written examination, you may be tested on:
1 how to use practical apparatus and techniques correctly
2 appropriate units for measurements
3 presenting your observations and data in an appropriate format

Practical activity groups (PAGs)

The specification lists a number of skills that you should have acquired and examples of apparatus that you should be able to use. These are given in Table 1.1.

Table 1.1 Practical activity groups

Practical activity group	Suitable practical activities
1 Microscopy (including staining and magnifications)	Study of the structure of plant, animal and prokaryotic cells and tissues Stages of mitosis, plasmolysis and crenation
2 Dissection (including drawing)	Dissection of a mammalian heart, mammalian kidney or plant stems
3 Sampling techniques	Measurement and calculation of species diversity (fieldwork)
4 Rates of enzyme-controlled reactions	Make quantitative measurements to study the effects of temperature, pH, substrate and enzyme concentration
5 Using a colorimeter or potometer to make quantitative measurements	Study of the effect of temperature on membrane permeability or the rate of an enzyme-catalysed reaction Investigation into the factors affecting the rate of transpiration
6 Chromatography or gel electrophoresis	Separation of photosynthetic pigments, a mixture of amino acids or DNA fragments
7 Aseptic microbiological techniques	Study of the effect of antibiotics on microbial growth
8 Transport in and out of cells	Investigation into diffusion, osmosis and active transport Estimation of the water potential of plant tissue such as potato tuber
9 Qualitative testing	Test for biological molecules (proteins, lipids, sugars and starch)
10 Investigation using ICT (data logger or computer modelling)	Use of a range of sensors such as temperature, pH or movement Investigation into DNA structure using RasMol
11 Investigation into the measurement of plant and animal responses	Investigation into tropism in plants, growth requirements of bacteria, human pulse rate at rest and after exercise, breathing rate and oxygen uptake by a human at rest and during exercise using a spirometer Use of *Drosophila* for genetic investigations
12 Research skills	Use of online and offline resources and citation of sources of information correctly, e.g. investigation into respiration in yeast

Presentation of results

Your results need to be presented in a table in the appropriate format:
1 There should be an informative title.
2 All raw data should be placed in a single table with ruled lines and a border.
3 Input or independent variable (IV) should be in the first column.
4 Output or dependent variable (DV) should be in columns to the right.
5 Processed data (e.g. means, rates, standard deviations) should be in columns to the far right.
5 There should be no calculations in the table, only calculated values.
6 Each column should have an informative heading with the correct SI units placed in brackets.
7 Units should appear only in the column headings, not in the body of the table.

Exam practice answers and quick quizzes at **www.hoddereducation.co.uk/myrevisionnotes**

8 All raw data should be recorded to the same number of decimal places and significant figures appropriate to the least accurate piece of equipment used to measure it.

9 All processed data may be recorded to one decimal place more than the raw data.

Analysis

As part of the written examination, you may need to:

1 process, analyse and interpret qualitative and quantitative experimental results

2 use appropriate mathematical skills for the analysis of quantitative data

3 use significant figures appropriately

4 plot and interpret suitable graphs from experimental results, including (i) selection and labelling of axes with appropriate scales, quantities and units, and (ii) measurement of gradients and intercepts

Processing

This involves calculating means, rates of reaction or some other aspect of the data. It also involves being able to draw a suitable graph. The type of graph used should be appropriate to the data collected. For all graphs:

- the diagram should be of an appropriate size to make good use of the paper and include an appropriate scale
- there should be an informative title
- the axes must be the correct way around (independent variable on the horizontal axis), labelled and with units in brackets

Bar charts

1 Used when the independent variable is discontinuous or non-numerical, e.g. colours, species.

2 Data presented as lines or blocks of equal width that do not touch.

3 Lines or blocks can be arranged in any order but could be ascending or descending.

Histograms

1 Used when the independent variable is continuous and numerical.

2 Blocks should be drawn touching.

3 The y-axis represents the number in each class or the frequency.

Pie charts

1 Used when displaying data that are proportions or percentages.

2 Should not contain more than six or seven sectors to avoid confusion

3 Segments should be labelled or a key provided.

Line graphs

1 Use a true origin (0,0) or, if the origin is not included, the axis should be broken.

2 Points should be plotted with saltire crosses (×) or encircled dots.

3 Straight lines should join the points or a smooth curve may be drawn if there is reason to believe that intermediate values fall on the curve.

4 If more than one curve is drawn on the same axes, each curve must be labelled to show what it represents.

Analysing

Use one of the statistical tests available. You should know when to use each of these tests:

1 the chi-squared test
2 a correlation coefficient
3 the Student's t-test

Use of significant figures

Any column in a table should have all the figures expressed to the same number of decimal points. The number of figures after the decimal point indicates precision in your measurements — this must be appropriate to the techniques or apparatus used. A mean or other processed data can be given to one extra figure after the decimal point.

Interpreting

1 Describe the trend shown in the data.
2 Make deductions and conclusions from your graphical data.
3 Justify these deductions by referring to the data.

Evaluation

REVISED ☐

You need to evaluate your results and draw conclusions. This includes:

1 identifying anomalies in experimental measurements
2 identifying limitations in experimental procedures
3 understanding the meaning of precision and accuracy in the measurement of results
4 appreciating the margin of error, percentage errors and uncertainties in apparatus
5 suggesting improvements to experimental design, procedures and apparatus

Identifying anomalies

An anomaly is a result that does not fit the pattern shown in your other measurements. It may be caused by:

1 a procedural error, such as diluting a solution incorrectly
2 a timing error, such as leaving the reaction to continue for longer than the other readings
3 misreading the scale on the apparatus
4 living material that does not behave consistently
5 not controlling some other factor that affects the results (this is called a limitation)

Including an anomaly in the calculated mean makes the mean less reliable.

Limitations

Limitations are factors that may affect the results but have not been controlled or accounted for in some way. Examples include:

1 small data sets
2 variables that cannot be controlled
3 the degree of precision of the instruments used
4 the accuracy of the instruments used
5 insufficient repeats
6 insufficient time for acclimatisation or equilibration before taking measurements

Exam practice answers and quick quizzes at **www.hoddereducation.co.uk/myrevisionnotes**

Precision and accuracy

Precision is the ability to be exact and it depends on the ability of the equipment used and the units of measurement. Accuracy is how close your measurement comes to the true value.

Reliability

Reliability is how much confidence you have in the results. This depends on the errors and uncertainty. Errors can be caused by:
1 unsuitable practical procedures
2 poor judgement by the experimenter

Uncertainty is a measure of the precision and accuracy of the apparatus used. The level of uncertainty can be taken as half the smallest unit used for measurement. If you are measuring length using a ruler marked in millimetres, the level of uncertainty is 0.5 mm. If you are using a digital scale, the level of uncertainty is one full unit.

Where a reading relies on the judgement of the experimenter, such as when a colour has changed, the level of uncertainty (and therefore the level of reliability) is much larger and hard to quantify.

Reliability can be assessed by making repeat measurements. This enables the experimenter to:
1 identify anomalies
2 calculate a mean
3 calculate standard deviation
4 calculate percentage error or percentage uncertainty:

$$\text{Percentage uncertainty} = \frac{\text{uncertainty}}{\text{actual measurement made}} \times 100$$

Using range bars and error bars

Your data can be used to assess the reliability of your results. Your measurements of the dependent variable will be spread around the mean.

The range is the difference between the highest and the lowest figures. The greater the range, the less reliable your results. For example, if you take three readings of the dependent variable (34, 36 and 32), the mean is:

$$\frac{34 + 36 + 32}{3} = 34$$

The range is 32 to 36.

Range bars are vertical bars drawn on your graph to show the range of your results for a particular value of the independent variable.

The standard deviation of your results is a measure of the spread of the results. It is given by the formula:

$$s = \sqrt{\frac{\sum(x - \bar{x})^2}{n-1}}$$

where s = standard deviation, $\bar{x}$ = the mean value and n = the number of data points. You can use the formula to calculate the standard deviation of your results. A larger standard deviation indicates a greater spread of results and less reliable data.

Error bars are drawn by plotting the values of the mean plus or minus the standard deviation (Figure 1.1).

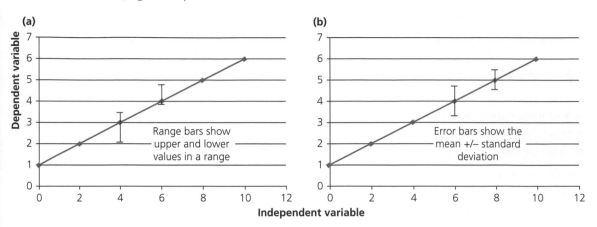

Figure 1.1 Graphs showing (a) range bars and (b) error bars

Summary

Practical skills will be tested in the written examination.

Planning may involve:
- solving a problem in a practical context
- selection of suitable apparatus, equipment or techniques
- describing a method or technique
- application of scientific knowledge
- identification of all variables
- evaluation of the method

Implementing may involve:
- describing how to use a wide range of practical apparatus and techniques
- suggesting precautions that should be taken to ensure results are valid
- assessing the hazards and potential risks involved
- deciding on appropriate units for measurements
- presentation of observations and data in an appropriate format (table)

Analysis may involve:
- processing qualitative and quantitative experimental results

- interpreting qualitative and quantitative experimental results
- using appropriate mathematical skills
- using significant figures appropriately
- plotting and interpreting suitable graphs from experimental results, including (i) selection and labelling of axes with appropriate scales, quantities and units, and (ii) measurement of gradients and intercepts
- drawing conclusions from data

Evaluation may involve:
- identifying anomalies in experimental measurements
- identifying limitations in experimental procedures
- understanding the meaning of precision and accuracy in the measurement of results
- appreciating the margin of error, percentage errors and uncertainties in apparatus
- calculating the standard deviation of a data set
- assessing the reliability of data using range bars and error bars
- calculating percentage error
- suggesting improvements to experimental design, procedures and apparatus
- assessing the validity of data

Exam practice

Paper 1 will contain some multiple-choice questions and short-answer structured questions. Paper 2 will contain short-answer structured questions and extended-response questions.

1 What is the most appropriate unit to measure the length of a mitochondrion? [1]
   A nm   **B** µm   **C** mm   **D** m

2 Results can be displayed in a number of different graphical forms. Which row in the following table correctly identifies the best way to display each set of data? [1]

Row	The effect of temperature on enzyme activity	A comparison of biodiversity in two fields	A belt transect across a ditch	A survey of the length of leaves on an oak tree
A	Histogram	Kite graph	Histogram	Bar chart
B	Line graph	Histogram	Kite graph	Pie chart
C	Line graph	Pie chart	Kite graph	Histogram
D	Line graph	Pie chart	Kite graph	Bar chart

3 A student investigated the effect of light intensity on the rate of transpiration from a leafy shoot using a potometer. Which row in the following table correctly identifies the variables? [1]

Row	Variables			
	Independent	Dependent	Control	Control
A	Light intensity	Rate of transpiration	Temperature	Air movement
B	Rate of transpiration	Light intensity	Temperature	Air movement
C	Light intensity	Air movement	Temperature	Rate of transpiration
D	Temperature	Rate of transpiration	Light intensity	Air movement

4 You are asked to make drawings from a slide of a plant leaf transverse section.
   (a) (i) What type of drawing should be made using a 4× objective lens? [1]
       (ii) Leaves contain palisade cells. List three other features you may include in your drawing. [3]
   (b) Using a 10× eyepiece and a 40× objective lens you could draw individual plant cells.
       (i) What is the total magnification used? [1]
       (ii) List three features that you may include in your drawing of a palisade cell. [3]

5 A student investigated the effect of temperature on the rate of decomposition of hydrogen peroxide by yeast. He collected the gas released in a measuring cylinder. The following table shows the results of his investigation.

Gas produced (cm³)			Temperature (°C)
Trial 1	Trial 2	Mean	
39	62	50.5	20
53	65	59	25
78	72	75	30
102	97	99.5	35

   (a) Outline how the student may have carried out this investigation. [5]
   (b) State two precautions that he should have taken to ensure the results were reliable. [2]
   (c) Identify three ways in which the table of results could be improved. [3]
   (d) Identify one anomalous result in the table and justify your choice. [3]
   (e) Suggest two limitations to this investigation. [2]
   (f) Suggest two ways in which the results collected could have been improved. [2]

Answers online

ONLINE

2 Cell structure

Microscopes and images

There is a range of microscopes for creating images of cells and ultrastructure. Each microscope has different limits of magnification and resolution.

Magnification and resolution

REVISED

Magnification is the ratio of image size to object size (size of image/size of object). **Resolution** is the ability to distinguish between two objects that are close together — the ability to provide detail in the image.

$$\text{Magnification } (M) = \frac{\text{image size } (I)}{\text{actual size } (A)}$$

Exam tip

Every biology examination paper includes a calculation. This may be to do with magnification or image size. You need to be able to calculate the magnification, but you also need to be able to manipulate the formulae.

Four types of microscope are commonly used, as shown in Table 2.1. Their advantages and disadvantages are given in Table 2.2.

Table 2.1 Types of microscope

Type of microscope	Magnification	Resolution	Use
Light	1000–2000×	50–200 nm	Viewing tissues and cells
Scanning electron	50000–500000×	0.4–20 nm	Viewing the surface of cells and organelles Providing depth/three-dimensional images
Transmission electron	300000–1000000×	0.05–1.0 nm	Detailing organelles (ultrastructure)
Laser scanning confocal	1000–2000×	50–200 nm	Three-dimensional images with good depth selection

Magnification is the ratio of the image size to the object size.

Resolution is the ability to distinguish two separate points that are close together.

Revision activity

Make up an acronym to help you remember the magnification formula.

Now test yourself

1 Using the formula $M = \frac{I}{A}$, rearrange the letters to give the formulae for I and for A.

Answer on p. 116

TESTED

Now test yourself

TESTED

2 Explain why a light microscope will not usually magnify images to greater than 1500×.
3 How can you tell the difference between an image created by a scanning electron microscope and one created by a transmission electron microscope?
4 Explain the relationship between millimetres (mm), micrometers (μm) and nanometers (nm).

Answers on p. 116

Table 2.2 Advantages and disadvantages of different microscopes

Type of microscope	Advantages	Disadvantages
Light	Cheap and easy to use Allows us to see living things	The resolution is limited
Scanning electron and transmission electron	The resolution is better than a light microscope, which means it is worth magnifying the image more as the image will show more detail Scanning electron microscopes also: ● give 3D images with depth of field ● are good for viewing surfaces	Electron microscopes are large and very expensive They require trained operatives The sample must be dried out and is therefore dead. This may affect the shape of the features seen (called an artefact) The image is in black and white, but colours may be added later by computer graphics. These are called false colour electron micrographs
Laser scanning confocal	Can also see living things and have the advantage that they can focus at a specific depth so the image is not confused by other components that are not in focus	Relies on a computer to piece together all the information from the dots of light created by the lasers. This means that the image is an interpretation rather than a real-life image

Staining

REVISED

Most cell components are colourless and hard to see. **Staining** is the application of coloured stains to the tissue or cells. Staining:

1 makes objects visible in light microscopes

2 increases contrast so that the object can be seen more clearly

3 is often specific to certain tissues or organelles. For example, acetic orcein stains chromosomes dark red, eosin stains cytoplasm, Sudan red stains lipids, and iodine in potassium iodide solution stains the cellulose in plant cell walls yellow and starch granules blue/black

In an electron microscope, the stains are actually heavy metals or similar atoms that reflect or absorb the electrons.

> **Revision activity**
>
> Draw a mind map to show the reasons for staining cells and tissues. Include the names of any stains you may have used such as methylene blue and iodine.

Cells

Cells are the basic unit of living organisms. All eukaryotic cells share a similar basic structure containing membrane-bound organelles. Each organelle, whether membrane-bound or not, has its own function within the cell.

Cell structure under the light microscope

REVISED

You should be able to prepare cells and tissues to view under a light microscope. This involves creating a smear or cutting a very thin section. These cells can then be stained appropriately and covered by a cover slip. Viewing these cells, interpreting what you can see and drawing what you see are important skills, as shown in the examples in Figure 2.1. Remember that details are important.

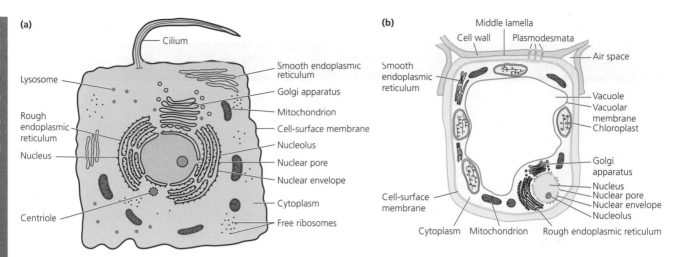

Figure 2.1 Diagrams of (a) an animal cell and (b) a plant cell as seen using an electron microscope

Cell ultrastructure

The **ultrastructure** of a cell is the detail you can see using an electron microscope (Table 2.3).

Table 2.3 Organelles and their functions

Organelle	Function	Diagram
Cell wall	Cellulose cell wall surrounds plant cells	
Centrioles	Involved in the organisation of the microtubules that make up the cytoskeleton Form the spindle fibres used to move chromosomes in nuclear division	
Chloroplasts	Site of photosynthesis	
Cilia	Small hair-like extension of cell-surface membrane containing microtubules Large numbers work in synchronised fashion Able to move whole organism or to move fluid (mucus) across a surface	
Cytoskeleton	A network of microtubules and microfilaments Provides mechanical strength, support and structure for the cell Maintains the cell shape and is used in some cells to change the shape of the cell Enables movement of organelles inside the cell Enables movement of the whole cell	

Exam practice answers and quick quizzes at **www.hoddereducation.co.uk/myrevisionnotes**

Organelle	Function	Diagram
Flagella (undulipodium in eukaryotes)	Large extension of cell-surface membrane containing microtubules (in eukaryotes) Able to beat to enable locomotion or move fluids	
Golgi apparatus	Modifies proteins made in the ribosomes Often adds a carbohydrate group Repackages proteins into vesicles for secretion	
Lysosomes	Small vacuoles containing hydrolytic or lytic enzymes	
Mitochondria (singular: mitochondrion)	Site of aerobic respiration	
Nucleus, nucleolus and nuclear envelope	Contains the genetic material (chromosomes) Controls the cell activities The nuclear envelope separates the genetic material from the cytoplasm The nuclear pores allow molecules of mRNA to pass from the nucleus to the ribosomes in the cytoplasm The nucleolus assembles the ribosomes	
Ribosomes	Site of protein synthesis	
Rough endoplasmic reticulum (RER)	Holds many of the ribosomes Provides a large surface area for protein synthesis	
Smooth endoplasmic reticulum (SER)	Associated with the synthesis, storage and transport of lipids and carbohydrates	

Now test yourself

TESTED

5 Explain why organelles such as mitochondria do not always look the same size and shape.

Answer on p. 116

Revision activity

From memory, make a list of all the membrane-bound organelles and note one function next to each. Make a separate list of the organelles that are not membrane-bound.

How the organelles work together

The **organelles** in a cell work together to achieve the overall function of that cell. Many of the organelles are involved in the production and secretion of **proteins**. The sequence of events always follows the same course:

1 mRNA leaves the nucleus via the nuclear pores.
2 It is used by the ribosomes on the rough endoplasmic reticulum to construct a protein.
3 The protein travels in a vesicle to the Golgi apparatus.
4 The vesicle is moved by the **cytoskeleton**, possibly using tiny protein motors that 'walk' along the microtubules using them as a track.
5 The Golgi apparatus modifies the protein (often adding a carbohydrate group) and repackages it into a vesicle.
6 This vesicle is moved to the cell-surface (plasma) membrane.
7 The vesicle fuses with the membrane to release the protein from the cell.

> **Revision activity**
>
> Draw a flow diagram of the sequence of events leading to the secretion of a protein.

> **Exam tip**
>
> Always remember to say that the plasma membrane is involved in secretion and that the vesicle fuses to this membrane.

Prokaryotic and eukaryotic cells

There are two types of cell: **prokaryote** (Figure 2.2) and **eukaryote** (Figure 2.1a). The features of each type are given in Table 2.4.

Table 2.4 Comparing prokaryotic and eukaryotic cells

Feature	Prokaryote	Eukaryote
Size	Smaller — typically less than 10 µm long and 1–2 µm wide	Larger — typically larger than 10 µm in diameter
Nucleus	No	Yes
Membrane-bound organelles	No	Yes
Ribosomes	Yes, 18 nm in size	Yes, 22 nm in size
Chromosomes	A single loop of DNA, no histones	DNA associated with proteins (histones)
Flagellum	Some cells have a flagellum. It has a very different structure	Some cells have a flagellum with 9 + 2 structure of microtubules

> **Revision activities**
>
> ● Make a list of the organelles found in a plant cell.
> ● Make a list of the organelles found in an animal cell.
> ● List the features of a bacterial cell that are also seen in plant cells.

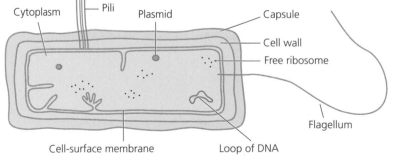

Figure 2.2 labels: Cytoplasm, Pili, Plasmid, Capsule, Cell wall, Free ribosome, Flagellum, Loop of DNA, Cell-surface membrane

Figure 2.2 The generalised structure of a prokaryotic cell

> **Revision activity**
>
> Write a list of all the key terms used in this chapter, then add the meaning of each key term.

Exam practice answers and quick quizzes at **www.hoddereducation.co.uk/myrevisionnotes**

Exam practice

Paper 1 will contain some multiple-choice questions and short-answer structured questions. Paper 2 will contain short-answer structured questions and extended-response questions.

1 Which row in the following table correctly identifies the sequence of cell components used during the secretion of a protein? [1]

Row	Sequence			
A	Golgi apparatus	Vesicles	Ribosomes	Plasma membrane
B	Vesicles	Ribosomes	Golgi apparatus	Plasma membrane
C	Plasma membrane	Vesicles	Golgi apparatus	Ribosomes
D	Ribosomes	Vesicles	Golgi apparatus	Plasma membrane
E	Golgi apparatus	Vesicles	Ribosomes	Plasma membrane

2 Which row in the following table correctly describes the components found in prokaryotic and eukaryotic cells? [1]

Row	Prokaryotic cells	Eukaryotic cells
A	Mitochondria, chloroplasts, ribosomes	Mitochondria, chloroplasts, ribosomes
B	Ribosomes, plasmids, flagellum	Nucleus, ribosomes, mitochondria
C	Mitochondria, chloroplasts, ribosomes	Nucleus, ribosomes, mitochondria
D	Nucleus, plasmids, ribosomes	Flagellum, chloroplasts, nucleus

3 The following statements are about the organelles in cells.
   A Mitochondria are where respiration occurs.
   B Chloroplasts contain chlorophyll.
   C Cilia are found only in the airways.
   D The cytoskeleton helps cells to change shape.
   E Ribosomes are larger than mitochondria.

   Which of the following options identifies the correct statements? [1]
   (a) All five statements are correct.
   (b) Only statements A, B and C are correct.
   (c) Only statements A, B and D are correct.
   (d) All five statements are incorrect.

4 Complete the following paragraph. [5]

   The four main types of microscope are ...., scanning electron, transmission electron and laser scanning confocal. The advantage of an electron microscope is that it has much greater .... than other types of microscope. This allows the user to see far more detail. A transmission electron microscope is used to see the detail of .... whereas a scanning electron microscope gives detail of the cell .... and a three-dimensional image. The advantage of a laser scanning confocal microscope is that it can produce images at a specific .... within the tissue or cell.

5 A student prepared a slide of onion epidermis using the stain methylene blue.
   (a) Explain the advantages of staining a tissue such as onion epidermis. [2]
   (b) The student drew one cell. Calculate the magnification of the image. [2]

10 µm

   (c) What feature shown in the student's diagram indicates that the cell is eukaryotic? [1]

6 (a) Complete the following table to show organelles and their functions. [4]

Organelle	Function
Mitochondrion	
Chloroplast	
	Modify proteins
	Manufacture proteins

(b) What is the role of the nuclear pores? [2]
(c) Describe how the structure of a mitochondrion is adapted to its function. [2]
(d) Suggest why lytic enzymes are held inside lysosomes. [2]
(e) Describe the roles of the cytoskeleton. [5]

Answers and quick quiz 2 online

ONLINE

Summary

By the end of this chapter you should be able to:
- Know the differences between light microscopes, electron microscopes and laser scanning confocal microscopes and describe the advantages and disadvantages of each.
- Appreciate the difference between resolution and magnification.
- Be able to calculate magnifications or true sizes given sufficient information.
- Understand that staining increases contrast and makes parts of the specimen stand out so that they can be seen more clearly.
- Differential staining allows different molecules, organelles or tissues to be stained different colours.

- Be able to recognise each type of organelle from a photomicrograph and from a drawing.
- Know what function each type of organelle has in the cell.
- Remember that most organelles consist of membranes and that the nucleus, the mitochondria and the chloroplasts have two membranes.
- Understand that membranes inside cells are distinct from the plasma (cell-surface) membrane that surrounds the cytoplasm.
- Be able to describe the sequence of events leading to the secretion of a protein from the cell.
- Know the differences between prokaryotic and eukaryotic cells.

3 Biological molecules

Water

Hydrogen bonds

Water is a simple molecule that can form **hydrogen bonds** between its molecules. This gives water some very important properties.

Hydrogen bonds are weak forces of attraction. They can form between molecules or between parts of a larger molecule. In water there is attraction between the oxygen of one molecule and the hydrogen of another molecule.

This attraction occurs because each water molecule is polar, which means there is an uneven distribution of charge. Oxygen atoms attract electrons more strongly than hydrogen atoms. Therefore, the electrons in a water molecule are pulled towards the oxygen atom, which gives the oxygen end of the molecule a more negative charge. However, the electrons are not pulled completely on to the oxygen atom, so the charge on the oxygen is not completely negative. This is shown as delta negative (δ^-). The hydrogen end of the molecule is left with a delta positive (δ^+) charge. It is these opposite delta charges that attract the water molecules together, producing hydrogen bonds (Figure 3.1).

> **Hydrogen bonds** are weak forces of attraction between molecules or parts of a molecule that are polar.

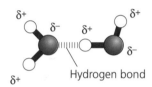

Hydrogen bond

Figure 3.1 Water molecules are polar and are held together by hydrogen bonds

Hydrogen bonds and the properties of water

Hydrogen bonds cause cohesion. Between 0°C and 100°C, they hold water molecules together loosely — they are held together, but they can move past one another and the water remains a liquid. In order to evaporate, the hydrogen bonds must be broken, allowing the molecules to separate and form a gas (water vapour). This takes a lot of energy, so water remains a liquid up to 100°C.

At lower temperatures the molecules have less kinetic (movement) energy and move about less. With less movement, more hydrogen bonds can form and at 0°C enough hydrogen bonds have formed to hold the water molecules in a stationary position, forming ice. The water molecules are now held in a rigid lattice, which holds the molecules further apart. Therefore, ice floats as it is less dense than water.

> **Exam tip**
>
> In any response, ensure you use the hydrogen bonds between the molecules of water to explain its properties.

> **Typical mistake**
>
> Many candidates fail to make the link between the action of the hydrogen bonds and the properties of water.

Water properties and living things

The following properties of water are important for the survival of many living things:
- Thermal stability — water has a high specific heat capacity, which means that a lot of energy is needed to warm it up. Therefore, a body of water maintains a fairly constant temperature, which is essential for life to survive.
- Freezing — ice is less dense than water so it floats, which insulates the water and prevents it freezing completely. Living things can survive below the ice.

> **Typical mistake**
>
> Many candidates fail to indicate how the properties of water are important to the survival of living things.

- Evaporation — a lot of energy is needed to cause evaporation, which is used to cool the surface of living things. This energy is known as latent heat. Water has a high specific latent heat capacity.
- At most temperatures water is a liquid — it can flow and transport materials in living things.
- **Cohesion** — the attraction of water molecules produces surface tension, which creates a habitat on the surface. It also enables continuous columns of water to be pulled up the xylem.
- Solvent — as the molecules are polar, water can dissolve a wide range of substances.
- As a reactant — water molecules are used in a wide range of reactions such as hydrolysis and photosynthesis.
- Incompressibility — water cannot be compressed into a smaller volume. This means it can be pressurised and pumped in transport systems or used for support in hydrostatic skeletons.

> **Cohesion** is the weak attraction between two similar molecules.

> **Revision activity**
>
> Draw a mind map with water in the centre to show how its properties are essential for living things.

Now test yourself

TESTED

1 Explain what is meant by a δ^+ charge and how it is created.
2 (a) Explain the difference between specific heat capacity and latent heat capacity.
  (b) What role does each play in the survival of living things?

Answers on p. 116

Monomers and polymers

The concept of monomers and polymers

REVISED

Many molecules are **polymers**. These are long chains that comprise a number of identical smaller molecules known as **monomers**. There are three biologically important groups of polymers found in living organisms: nucleic acids, polysaccharides and proteins.

In nucleic acids, the monomers are nucleotides. They are made up of a:
1 five-carbon or pentose sugar. This is either ribose (in the case of RNA) or deoxyribose (in DNA)
2 phosphate group
3 nucleotide base. There are five different bases. These are adenine, cytosine, guanine, thymine and uracil, which are usually abbreviated to A, C, G, T and U. The nucleotides that make up DNA contain one of the bases adenine, cytosine, guanine or thymine. In RNA, the thymine is replaced by uracil

Complex carbohydrates such as starch and cellulose are polysaccharides — a group of polymers made up of monosaccharides.

In proteins, the monomers are amino acids. There are 20 different amino acids used to build proteins. The different sequences that these amino acids can be used in provide huge diversity in protein structure.

> **Polymers** are long chains of repeated units. The individual units are called monomers.

> **Monomers** are one of the small similar molecules that join together to form a polymer. There are three important biological monomers: amino acids, monosaccharides and nucleotides.

Condensation and hydrolysis

REVISED

The monomers in both complex carbohydrates and proteins are joined by covalent bonds created by a **condensation** reaction (Figure 3.2). This is a reaction that releases a water molecule. The bond can be broken again by adding a water molecule. This is called **hydrolysis**.

> **Condensation** is a reaction that releases a molecule of water.

Exam practice answers and quick quizzes at **www.hoddereducation.co.uk/myrevisionnotes**

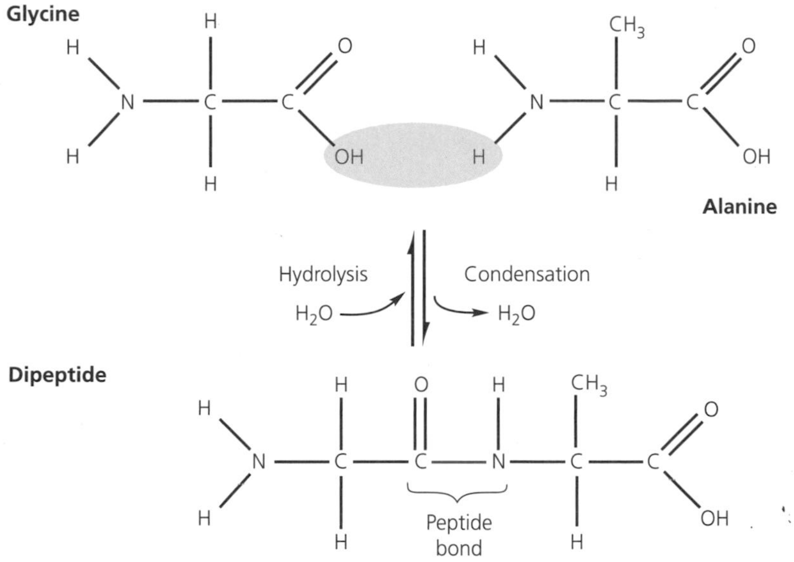

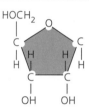

Figure 3.2 Condensation forms a peptide bond between two amino acids. Hydrolysis can break the bond

> **Hydrolysis** is splitting a large molecule into two smaller molecules by the addition of water.

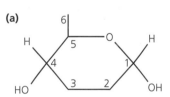

The chemical elements that make up biological molecules

Carbohydrates

Carbohydrates consist of carbon, oxygen and hydrogen. The general formula for a carbohydrate is $C_x(H_2O)_x$. Carbohydrates can be divided into three main groups: monosaccharides, disaccharides and polysaccharides.

Monosaccharides

These are the single sugar units that are used as monomers to build other carbohydrates. They are soluble and sweet reducing sugars.

Pentose sugars

Pentose monosaccharides contain five carbons, e.g. **ribose** and deoxyribose (Figure 3.3).

Figure 3.3 Ribose, a pentose sugar

Hexose sugars

Hexose monosaccharides contain six carbons, e.g. **glucose**. There are two forms of glucose: alpha (α) glucose and beta (β) glucose. The structures of α-glucose and β-glucose are shown in Figure 3.4.

The difference between α-glucose and β-glucose is simply the position of the −H and −OH groups on the first carbon atom. This difference is not great, but it has a large effect on the way they bond together and the polysaccharides produced.

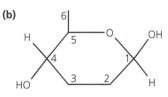

(a) α-glucose (note that the −H is above the −OH on carbon atom 1)

(b) β-glucose (note that the −OH is above the −H on carbon atom 1)

Figure 3.4 (a) α-glucose and **(b)** β-glucose. Note that the carbon atoms are numbered from the oxygen clockwise

Disaccharides

Disaccharides are two monosaccharides bonded together. They are bonded by a covalent bond called a **glycosidic bond**. Disaccharides are soluble and sweet. Most are reducing sugars, but sucrose is not a reducing sugar.

The **glycosidic bond** is the bond between two monosaccharides in a polysaccharide.

Forming disaccharides

Disaccharides are formed by a condensation reaction between two monosaccharides. The most common bond is between the carbon 1 of one monosaccharide and the carbon 4 of another. This forms a 1,4 glycosidic bond and releases water.

Glucose and fructose combine to form sucrose, whereas glucose and glucose combine to form maltose (Figure 3.5), and glucose and galactose combine to form lactose. Disaccharides can be converted back to monosccharides by hydrolysis.

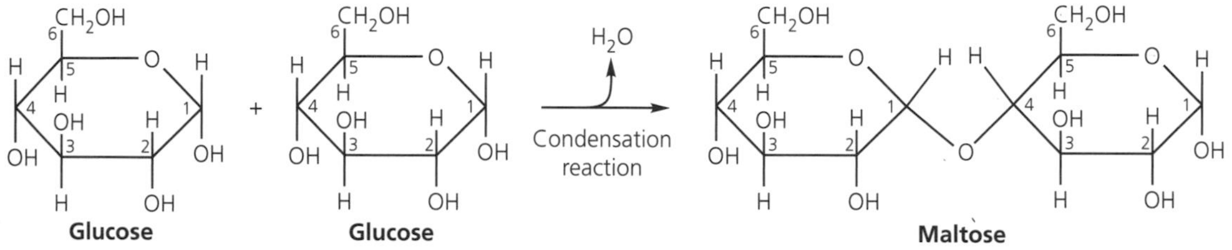

Figure 3.5 Two α-glucose molecules combine to form maltose

Polysaccharides

Polysaccharides are large insoluble molecules. **Starch** is actually a combination of two molecules, amylose and amylopectin.

Amylose

Amylose consists of a long unbranched chain of α-glucose subunits. The subunits are joined by 1,4 glycosidic bonds. The chain of subunits coils up. The hydroxyl group on carbon 2 of each subunit is hidden inside the coil. This makes the molecule less soluble.

Amylose is used for the storage of glucose subunits and energy in plant cells. The molecule is compact — it takes little space in the cell. It is insoluble, which means the molecules do not affect the water potential of the cells.

Glucose subunits can be removed easily from each end of the molecule. They can be used as building blocks to build other substances or as a substrate in respiration to release stored energy.

Amylopectin and glycogen

Amylopectin and **glycogen** are similar to amylose in that they are both long chains of α-glucose subunits bonded by 1,4 glycosidic bonds (Figure 3.6). Some of the glucose subunits also have 1,6 glycosidic bonds as well as the 1,4 glycosidic bonds. This means that the molecule is branched.

Amylopectin occurs in plants and has few branches. Glycogen is used for storage of glucose subunits and energy in animal cells. Glycogen has more 1,6 glycosidic bonds, making it more branched. This means that the molecule has many ends from which glucose can be released quickly. It has the same advantages as amylose: it is insoluble and compact.

Exam tip

Complex carbohydrates are not that complex — they are simply long chains of one or two subunits. Each chain is called a polymer.

Exam tip

You need to know about amylose and amylopectin, but you should be able to apply that knowledge to other complex carbohydrates.

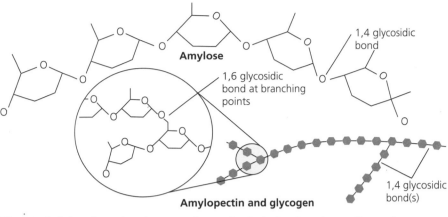

Figure 3.6 Amylose is a long, unbranched chain. Amylopectin and glycogen have branches formed by 1,6 glycosidic bonds

Cellulose

Cellulose consists of a long unbranched chain of β-glucose subunits. The subunits are joined by a 1,4 glycosidic bond. The chain of β-glucose subunits form a straight chain (Figure 3.7).

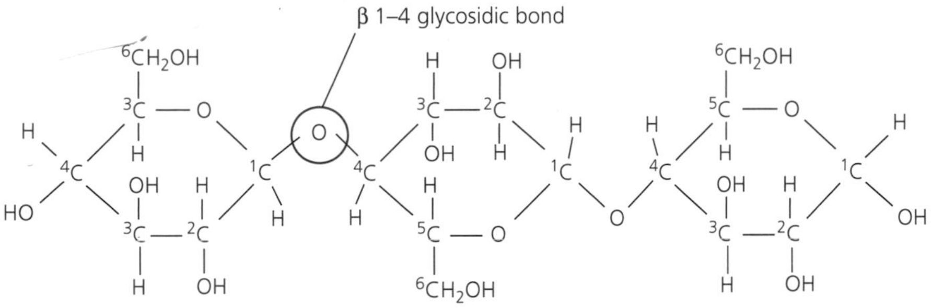

Figure 3.7

The hydroxyl groups on carbon 2 of each subunit are exposed, allowing hydrogen bonds to form between adjacent cellulose molecules. Some 60–70 molecules bind together to form a cellulose microfibril and many microfibrils join together to form macrofibrils.

Cellulose is strong and completely insoluble. It is used in plant cell walls and provides enough strength to support the whole plant.

Now test yourself

TESTED

4 Explain why cellulose is insoluble.

Answer on p. 116

Lipids

REVISED

Lipids are not polymers like proteins and complex carbohydrates. They are a large group of compounds that includes triglycerides, phospholipids and steroids. Lipids are insoluble in water.

Triglycerides

A **triglyceride** is a **macromolecule** containing one glycerol molecule and three fatty acid chains (Figure 3.8). The fatty acids are attached to the glycerol by a condensation reaction. The bonds are called **ester bonds** and they can be broken by hydrolysis.

Now test yourself

3 List the reasons why amylose is a good storage product.

Answer on p. 116

TESTED

Typical mistake

Many candidates are mistaken in thinking that cellulose is a protein, not a carbohydrate.

Exam tip

The structure of complex carbohydrates lends itself to a question in which the examiner asks you to relate the structure of the molecule to the function of that molecule. This is perhaps most easily done as a table.

Revision activity

Draw a table to compare the structures and properties of amylose, glycogen and cellulose.

A **triglyceride** is a molecule that comprises one glycerol molecule and three fatty acid chains.

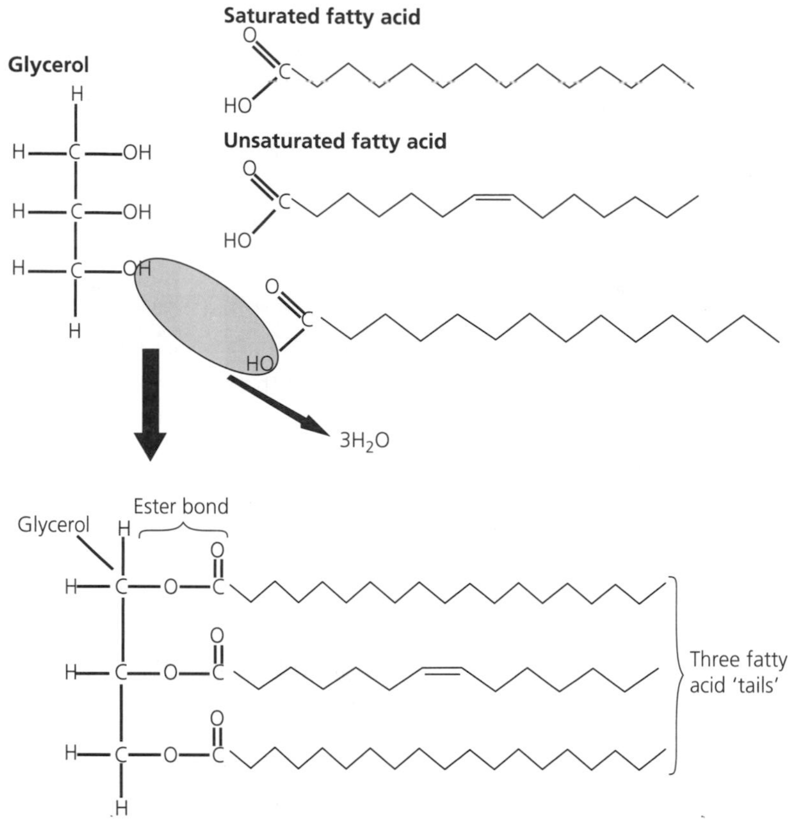

Figure 3.8 Three fatty acids combine with one glycerol to produce a triglyceride. The elimination of one water molecule is shown. Three water molecules are released

Triglyceride molecules are rich in energy and used to store excess energy. When required, the molecules can be broken down in aerobic respiration to release this energy. Water is also released, which can be useful for animals that live in dry environments — hence camels store fat in their humps. The stores can be held under the skin and around major organs. It has the benefit of protecting the major organs from physical shock.

Triglycerides are also good insulators and are used to insulate animals that live in cold environments such as polar bears and aquatic mammals such as whales. They also provide buoyancy for these mammals.

Saturated and unsaturated fatty acids

Fatty acids are long chains of carbon atoms with hydrogen atoms bonded to them. If each carbon has two hydrogen atoms attached, there are no double or triple bonds in the fatty acid. This is called a saturated fatty acid. Saturated fatty acids are found in animal fats. They have a higher melting point and are more solid at room temperature, like butter.

If there are fewer hydrogen atoms, there will be double or even triple bonds between adjacent carbon atoms. This is called an unsaturated fatty acid. Unsaturated fatty acids are found in plant fats and oils. They have lower melting points and are more likely to be liquid at room temperature, like spreads and vegetable oil.

Now test yourself

5 Explain why lipids are good storage molecules.

Answer on p. 116

TESTED

Phospholipids

Phospholipids are similar to triglycerides, but one of the fatty acid chains is replaced by a phosphate group (Figure 3.9). The two remaining fatty acid 'tails' are insoluble in water and are called **hydrophobic**.

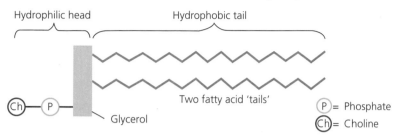

Figure 3.9 A phospholipid has two fatty acid chains and a phosphate group

The phosphate group is complex and includes choline, which is water soluble. This alters the characteristics of the molecule. This group makes the 'head' end of the phospholipid able to mix with water — it is **hydrophilic**. Phospholipids form bilayers with the hydrophobic 'tails' in the centre and the hydrophilic 'heads' pointing outwards to interact with the surrounding aqueous solution. This is the basis of all cell membranes.

> A **phospholipid** is a molecule containing one glycerol molecule, a phosphate group and two fatty acid chains.
>
> **Hydrophobic** means water-hating, repelled by water.

> **Hydrophilic** means water-loving, attracted to water.

Now test yourself

TESTED

6 Explain why phospholipids are used in membranes.

Answer on p. 116

> **Revision activity**
>
> Write an exam-style question worth about 10 marks about the roles of lipids in living things. Ensure that at least 5 marks test assessment objective 2 (use of your knowledge in an unfamiliar context). Write out a mark scheme.

Amino acids and proteins

REVISED

Amino acids

Proteins are made up of long chains of **amino acids**. There are 20 different amino acids used in proteins, but all have the same basic structure (Figure 3.10). The residual *R* group is the only part that differs between different amino acids.

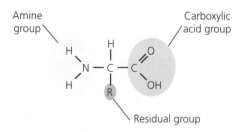

Figure 3.10 The generalised structure of an amino acid

Proteins

All proteins consist of long, unbranched chains of amino acids, which are held together by **peptide bonds**. These bonds are formed by **condensation** and occur between the amino group of one amino acid and the carboxylic acid group of another. A peptide bond is formed by condensation.

Two amino acids together make a **dipeptide** (Figure 3.11). Many amino acids in a chain form a **polypeptide**.

> A **peptide bond** is the bond between two amino acids.
>
> **Condensation** is a reaction that involves the release of water molecules.
>
> A **dipeptide** is formed when two amino acids are joined together by a peptide bond.
>
> A **polypeptide** is a chain of many amino acids joined together by peptide bonds.

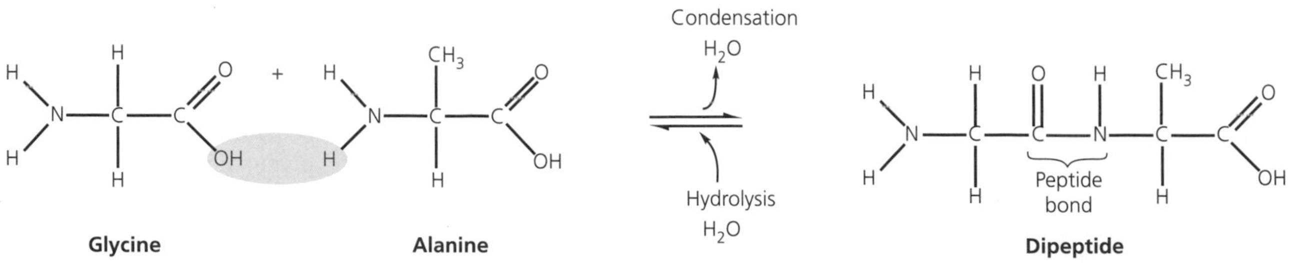

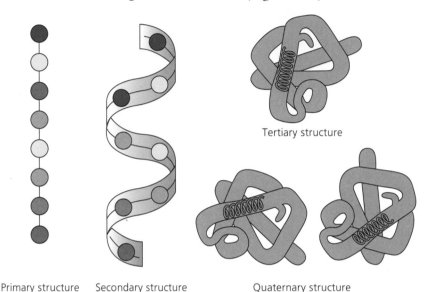

Figure 3.11 The condensation reaction to form a dipeptide

There are four levels of **protein structure** (Figure 3.12).

Now test yourself

7 Explain why proteins are unbranched chains.

Answer on p. 116

TESTED

Tertiary structure

Primary structure   Secondary structure   Quaternary structure

Figure 3.12 The four levels of structure in a protein

A **primary structure** is a chain of amino acids, held together by peptide bonds.

A **secondary structure** is formed when the chain of amino acids becomes folded and coiled. Two shapes are formed:
- alpha (α) helix — shaped like a coil spring
- beta (β) sheets — pleated like a folded sheet of paper

Hydrogen bonds hold the folds and coils in place.

A **tertiary structure** is formed when the coiled and pleated chains can be folded further to produce the final three-dimensional shape of the molecule. These final folds and coils are caused by the interactions between the R groups on the amino acids, which interact to form a range of bonds that hold the three-dimensional shape. These bonds include:
- hydrogen bonds between polar R groups
- ionic bonds between R groups with opposite charges
- covalent disulfide bonds between two sulfur-containing R groups

Additionally, some R groups are hydrophobic and twist away from water into the centre of the molecule. Others are hydrophilic and twist outwards so that they are on the outside of the molecule.

Many proteins are just one polypeptide chain that has been coiled and folded. However, some proteins consist of more than one polypeptide chain — e.g. haemoglobin has four polypeptide chains and collagen has three. These multi subunit proteins make up the **quaternary structure**.

Typical mistake

Some candidates suggest that a protein shows just one of the levels of structure or that the labelling applies to the nutritional value of the protein. This is not the case — all proteins show a primary, a secondary and a tertiary level of structure, and some show a quaternary level of structure.

Exam tip

A typical question might ask you to relate the structure of a protein to its properties — this could be answered in the form of a table.

Globular proteins

Globular proteins are proteins that are highly folded to form a globular shape. These proteins are more water soluble. They are active in metabolism and their activity relies on their three-dimensional shape. Their shape and activity are sensitive to temperature changes: higher temperatures can cause distortion of their shape.

Haemoglobin

Haemoglobin is a **conjugated protein**. It is used to transport oxygen in the form of oxyhaemoglobin. Haemoglobin contains four polypeptide chains called subunits — two alpha (α) chains and two beta (β) chains.

Haemoglobin is a conjugated protein as each subunit has a non-protein prosthetic group attached, called a haem group, which contains a single iron ion (Fe^{2+}). One oxygen molecule can attach to each haem group, so a haemoglobin molecule can carry four oxygen molecules.

Enzymes

An example of an **enzyme** is amylase, which hydrolyses the bonds between glucose subunits in amylose. The molecule has regions that are coiled in an alpha (α) helix and other regions that are folded into beta (β) sheets. The compact globular shape contains an active site that has a specific shape that is complementary to the shape of the substrate (in this case, amylose). The active site holds at least one calcium ion that acts as a cofactor — it is essential for the correct action of the enzyme.

Hormones

An example of a hormone is **insulin**, which is used to stimulate removal of excess glucose from the blood. There are two polypeptides held together by disulfide bridges. One polypeptide is 21 amino acids and the second is 30 amino acids. The molecule has a specific three-dimensional shape that is complementary to the shape of a glycoprotein receptor on the surface of cells in the liver.

Fibrous proteins

Fibrous proteins tend to have a regular sequence of amino acids that is repeated many times. They are less soluble in water (usually totally insoluble). They tend to form fibres that have structural functions and examples include collagen, keratin and elastin.

Collagen

Collagen has three polypeptide chains wound around one another. It is not easily stretched. It provides strength in the walls of arteries to withstand the high blood pressure. It is found in tendons, which hold muscle to bone, and in bone, where it is hardened by calcium phosphate.

Keratin

Keratin has two polypeptide chains coiled together. It is strong and is used for protecting delicate parts of the body. Examples include finger nails, claws, hooves, horns, scales, hair and feathers. The cells in the outer layer of skin also contain keratin, which makes them impermeable to water.

Now test yourself

8 Explain the difference between a polypeptide and a protein.

Answer on p. 116   TESTED

A **conjugated protein** is a globular protein with a prosthetic group.

Revision activity

Make a model (or draw a diagram) of a protein using coiled wire or a slinky spring and folded paper. Label and annotate the model to show examples of the different types of bonding.

3 Biological molecules

Elastin

Elastin is produced by linking many tropoelastin fibres together. Tropoelastin is coiled like a spring and can stretch and recoil. It is used wherever stretching and recoil is required, such as in the walls of the arteries and airways, alveoli, skin and the wall of the bladder.

Now test yourself

9 Explain why proteins that are metabolically active such as enzymes or hormones are globular proteins.
10 Explain why globular proteins are more affected by temperature change than fibrous proteins.

Answers on p. 116

The key inorganic ions involved in biological processes

Inorganic ions

REVISED

Inorganic ions are charged particles that have a number of important roles. These roles range from creating skeletal structures to nervous conduction and activating enzymes. Cations are positively charged and anions are negatively charged. Table 3.1 summarises these roles.

Table 3.1 The roles of inorganic ions

Ion	Role in biological systems
Ammonium (NH_4^+)	A component of amino acids, proteins and nucleic acids Involved in the: ● nitrogen cycle ● maintenance of pH
Calcium (Ca^{2+})	Increases the hardness of bones, teeth and the exoskeletons of crustaceans. It is also found in the middle lamella between plant cells A factor in blood clotting Involved in the control of muscle contraction and synaptic action Activates enzymes such as amylase and lipase
Chloride (Cl^-)	Involved in the: ● reabsorption of water in the kidney ● regulation of water potentials in cells and body fluids ● transport of carbon dioxide in the blood ● production of hydrochloric acid in the stomach
Hydrogen (H^+)	Involved in: ● oxidative phosphorylation in respiration ● photophosphorylation in photosynthesis ● the transport of carbon dioxide in the blood ● the regulation of blood pH ● reduction reactions in metabolism
Hydrogencarbonate (HCO_3^-)	Involved in the: ● regulation of blood pH ● transport of carbon dioxide in the blood

Ion	Role in biological systems
Hydroxide (OH^-)	Involved in the regulation of blood pH
Iron (Fe^{2+})	Increases the affinity of haemoglobin for oxygen
Sodium (Na^+)	Involved in: ● the regulation of water potentials in cells and body fluids ● the selective reabsorption of sugars and amino acids in the kidney ● the reabsorption of water in the kidney ● nervous transmission and muscle contraction
Magnesium (Mg^{2+})	Found at the centre of chlorophyll
Nitrate (NO_3^-)	A component of amino acids, proteins and nucleic acids Involved in the nitrogen cycle
Phosphate (PO_4^{3-})	A component of phospholipids, ATP and nucleic acids Increases the hardness of bones, teeth and the exoskeletons of crustaceans Improves root growth in plants
Potassium (K^+)	Improves growth of leaves and flowers in plants Involved in the: ● regulation of water potentials in cells and body fluids ● nervous transmission and muscle contraction

Testing for the presence of biological molecules

Chemical tests

REVISED

The presence of each biological molecule can be detected by a specific test. Table 3.2 summarises the procedures to be followed.

Table 3.2 Chemical tests to identify the presence of biological molecules

Molecule tested for	Test name	Details of test	Positive result
Protein	Biuret test	Dissolve in water. Add biuret A and biuret B	Colour change from blue to mauve or purple
Reducing sugar (e.g. glucose)	Benedict's test	Dissolve in water. Add Benedict's reagent. Heat at 80–90°C for 2 minutes	A precipitate forms. Colour change from blue to brick red
Non-reducing sugar (e.g. sucrose)	Benedict's test (this test converts non-reducing sugars to reducing sugars and then tests the reducing sugars)	Test the substance for reducing sugars to ensure that none are present, then dissolve in water. Add a few drops of dilute hydrochloric acid and boil for 2 minutes. Neutralise the solution by adding a few drops of dilute sodium hydroxide. Add Benedict's reagent and reheat for 2 minutes	A precipitate forms and there is a colour change from blue to brick red
Starch (amylose)	Iodine solution test	Dissolve in water. Add iodine solution	Colour change to deep blue/black
Fat (lipids)	Emulsion test	Dissolve in alcohol. Filter. Add water to the filtrate	When water is added to the clear filtrate, it will turn cloudy or milky

Making the Benedict's test quantitative

When testing for reducing sugars such as glucose, the colour change may not be complete: the colour may show a change from blue to green or yellow before going orange or red. If only a small amount of sugar is present, it will not react all the Benedict's reagent — leaving some of it blue. This is what causes an incomplete colour change.

The **Benedict's test** can be made quantitative (i.e. you can determine the concentration of the reducing sugars) by ensuring there is excess Benedict's reagent. Create a range of colours using known concentrations of reducing sugars to create a set of colour standards.

To make the measurement more precise, the standards and the sample should be placed in a centrifuge and spun for 2 minutes. This will deposit the coloured precipitate at the bottom of the tube, leaving a blue solution of unused Benedict's reagent. The more concentrated the reducing sugar, the less blue colour will be left. The intensity of the blue colour in the solution can be measured using a **colorimeter**. Plot a graph of absorbance against concentration using the standard solutions. Use your graph to determine the concentration of the unknown solution by reading from the absorbance measurement across to the concentration.

Now test yourself

TESTED

11 Explain why carrying out the Benedict's test on a solution of glucose at low concentration will leave the solution blue or green in colour, but if the glucose is at high concentration the solution will have no sign of blue colouration.

Answer on p. 116

Biosensors

A **biosensor** converts a chemical variable into an electrical signal. For example, a glucose biosensor measures the concentration of glucose. When the biosensor is dipped into a solution, the glucose diffuses towards immobilised enzymes. These catalyse a reaction that releases hydrogen peroxide. The hydrogen peroxide reacts with a platinum electrode to generate a current. The current generated is proportional to the glucose concentration.

Chromatography

Chromatography is a technique used to separate molecules in a mixture.

Paper chromatography involves placing a small sample of the mixture solution on to a strip of chromatography paper. The paper is then placed with one end in a shallow layer of solvent. The solvent rises up through the paper by capillary action. When it passes the sample of mixture, the molecules in the mixture start to travel up the paper with the solvent. Non-polar substances move more quickly up the paper strip.

Thin-layer chromatography (TLC) works in a similar way to paper chromatography.

R_f values

The R_f (retention factor) value is a measure of the distance moved by a particular molecule:

$$R_f = \frac{\text{distance moved by molecule}}{\text{distance moved by solvent front}}$$

The R_f value is specific to the molecule, the solvent and the type of paper or thin layer used for separating the mixture. It can be used to identify a molecule in the mixture.

> **Revision activity**
>
> Write a list of all the key terms used in this chapter, then add the meaning of each key term.

Exam practice

1 (a) List three functions of carbohydrates in living things. [3]
   (b) Describe and explain how the structure of starch (amylose) makes it suitable as a storage compound. [6]
   (c) Complete the following table to compare glycogen with cellulose. [5]

Structural feature	Glycogen	Cellulose
Sugar(s) present		
Bonds present		
Branched or unbranched		
Coiled or straight		
Forms cross-links with other molecules		

2 (a) (i) Explain what is meant by the term *primary structure of a protein*. [1]
       (ii) Explain what is meant by the term *secondary structure of a protein*. [3]
   (b) Describe the bonds involved in holding the tertiary structure of a protein. [4]
3 (a) Explain why hydrogen bonds form between water molecules. [3]
   (b) Explain the role of hydrogen bonds in making water a suitable medium for transport in living things. [10]
4 A student carried out a test to determine if a certain substance was present in the seeds of a small plant. She crushed the seeds and added some alcohol. She then filtered the solution before adding water to the filtrate. The resulting solution turned milky.
   (a) State what substance was tested for. [1]
   (b) Why was it necessary to filter the solution? [2]
   (c) Suggest how the student could make this test quantitative. [6]

Answers and quick quiz 3 online

ONLINE

Summary

By the end of this chapter you should be able to:
- Describe how hydrogen bonding occurs.
- Relate the properties of water to its roles in living things.
- Describe the structure of α-glucose and β-glucose and the formation of glycosidic bonds.
- Describe and compare the structures of amylose, glycogen and cellulose and explain how the structure of each molecule relates to its function.
- Describe and compare the structures of a triglyceride and a phospholipid and explain how the structure of each molecule relates to its function.
- Describe the structure of amino acids and the formation of peptide bonds.
- Explain the four levels of protein structure with reference to the types of bonds involved at each level.
- Describe and compare the structures of haemoglobin, an enzyme and a hormone as examples of a globular proteins.
- Understand the properties and functions of fibrous proteins.
- Know the functions of a range of inorganic ions in living things.
- Describe the chemical tests for protein, reducing sugar, non-reducing sugar, starch and lipid.
- Describe how tests can be made quantitative.
- Understand the principles of chromatography.

4 Nucleotides and nucleic acids

The structure of nucleotides and nucleic acids

RNA and DNA

Ribonucleic acid (RNA) and **deoxyribonucleic acid (DNA)** are **nucleic acids** made from **nucleotides** (Figure 4.1). A nucleotide has three components:
- a phosphate group
- a pentose sugar
- an organic base

Table 4.1 outlines the differences between the nucleotides in RNA and those in DNA.

Table 4.1 The differences between the nucleotides in RNA and DNA

Feature	RNA	DNA
Pentose sugar	Ribose	Deoxyribose
Purines (two rings)	Adenine and guanine	Adenine and guanine
Pyrimidines (one ring)	Cytosine and uracil	Thymine and cytosine

The structure of RNA

RNA is a **polynucleotide**. It is usually single-stranded and much shorter than DNA. The sugar involved is ribose, not deoxyribose. The bases used include adenine, cytosine, guanine and uracil. There are three types of RNA:
1 **messenger RNA (mRNA)**, which carries the code held in the genes to the ribosomes where the code is used to manufacture proteins
2 **transfer RNA (tRNA)**, which transports amino acids to the ribosomes
3 **ribosomal RNA (rRNA)**, which makes up the ribosome

The structure of DNA

DNA is also a polynucleotide. The organic bases can pair up — one purine with one pyrimidine. They pair according to their complementary shapes. Adenine always pairs to thymine (or uracil in RNA) using two hydrogen bonds. Cytosine always pairs to guanine using three **hydrogen bonds**. The two polynucleotide strands lie in opposite directions, which is known as **antiparallel**. The two single polynucleotide strands are joined together to make a double strand. The whole molecule twists to form a helix, hence the name **'double helix'** (Figure 4.2).

> **Nucleic acids** are complex organic substances present in living cells DNA or RNA molecules consist of many nucleotides linked in a long chain.
>
> **Nucleotides** are **monomers** from which nucleic acids are formed. They are a combination of a phosphate, a pentose sugar and an organic base.

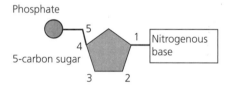

Figure 4.1 All nucleotides have the same structure. Note that the phosphate is attached to the fifth carbon atom in the sugar

> **Typical mistake**
>
> Candidates sometimes forget that the sugar in RNA is ribose not deoxyribose, as found in DNA.

> **Exam tip**
>
> Make sure that you know the key differences between RNA and DNA.

> **Revision activity**
>
> Draw a table to compare the structures of RNA and DNA.

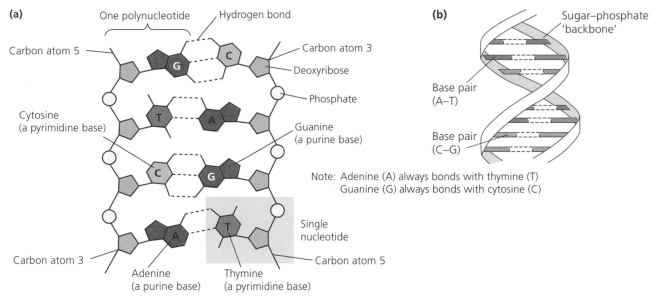

(a)

(b)

Note: Adenine (A) always bonds with thymine (T)
Guanine (G) always bonds with cytosine (C)

Figure 4.2 The structure of DNA showing (a) the hydrogen bonds and (b) the double helix

Typical mistake

Many candidates don't supply the detail about the number of hydrogen bonds used: adenine binds to thymine with two hydrogen bonds whereas cytosine bonds to guanine with three. This is why adenine cannot bind to cytosine and thymine cannot bind to guanine.

Polynucleotides

REVISED

A polynucleotide is formed when nucleotides bind together in a long chain (Figure 4.3). The bonds are formed by condensation and are called **phosphodiester bonds** (Figure 4.4). They can be broken by hydrolysis. These bonds form between the sugar of one nucleotide and the phosphate group of another, making a sugar–phosphate 'backbone'. This leaves the organic base of each nucleotide sticking out to the side of the chain.

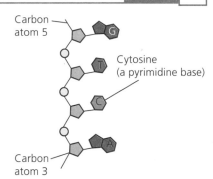

Figure 4.3 A polynucleotide

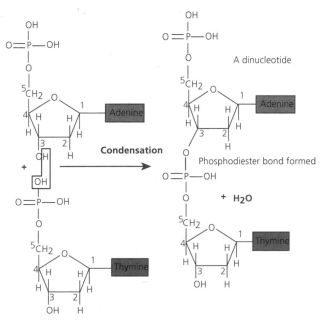

Figure 4.4 The formation of phosphodiester bonds

ADP and ATP

REVISED

ADP and **ATP** are **phosphorylated nucleotides**. They contain a **pentose sugar** (**ribose**), a **nitrogenous base** (**adenine**) and two or three **inorganic phosphates** (Figure 4.5).

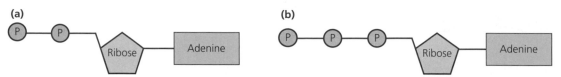

Figure 4.5 (a) ADP and (b) ATP are phosphorylated nucleotides

The replication of DNA

Semi-conservative DNA replication

REVISED

DNA **replication** occurs in all living organisms in order to copy their DNA for biological inheritance. Precise replication can take place because of the double-stranded structure of the DNA molecule. The process, known as **semi-conservative replication**, is as follows:

1 One double-stranded molecule untwists and the hydrogen bonds between the base pairs break. This is catalysed by the enzyme **helicase**.
2 The two polynucleotide chains separate, exposing the bases.
3 Each strand is then used as a template to make two new double strands.
4 New nucleotides pair to the exposed bases on both strands, using their complementary shapes to pair correctly.
5 The two new chains of nucleotides bond together by the enzyme **DNA polymerase** to form the second half of the DNA molecule.
6 The enzyme also checks that the pairing of the bases is correct.
7 Each new molecule then twists to form its double helix.

> **Replication** means to make an identical copy.
>
> **Semi-conservative replication** means that half of the original molecule is conserved in each of the new molecules.

Mutations

The precise replication of DNA is essential to ensure that identical copies of the genes are included in every cell of the body. Usually the precise pairing of the bases ensures that two exact copies of the original DNA molecule are made. However, occasionally an incorrect base may be bonded into place, which is known as a **mutation**. Mutations are **random** and **spontaneous**.

> **Revision activity**
>
> Draw a flow diagram to show the sequence of events in DNA replication.

Now test yourself

TESTED

1 Explain why it is important that each new double helix is identical to the original.

Answer on p. 117

The nature of the genetic code

The genetic code

REVISED

The **genetic code** is **universal** — all organisms use the same code. It is a sequence of bases. Three consecutive bases (known as **triplets**) codes for one amino acid. There are 64 possible triplet codes but only

20 amino acids used to make proteins. Some amino acids are coded for by more than one triplet (called **degenerate** codons). The triplets are **non-overlapping**. A particular sequence of triplets will code for one sequence of amino acids. A sequence of amino acids forms one polypeptide. The length of DNA that codes for one polypeptide is called a **gene**. One gene codes for one polypeptide. This may form a protein. If the protein contains more than one polypeptide (i.e. it has a quaternary structure), there will be more than one gene used to code for the protein.

The synthesis of polypeptides

The **synthesis of polypeptides** involves two stages:
1 **Transcription** — reading the code and producing a messenger molecule to carry the code out to the cytoplasm.
2 **Translation** — converting the code to a sequence of amino acids.

Transcription

The genetic code is held in the nucleus. The DNA molecule is too large to leave the nucleus, so a smaller messenger molecule is made called messenger RNA (mRNA). The double-stranded DNA molecule is unwound and split by the action of the enzyme **RNA polymerase**, which breaks the hydrogen bonds holding the two strands together. This exposes the sequence of bases in the gene. The coding strand carries the genetic code and the complementary strand is a non-coding strand sometimes called the antisense strand or the template strand, which is used to build a copy of the coding strand called messenger RNA (mRNA). It has an identical sequence of bases to the coding strand of the gene, except that the thymine is replaced by uracil. Each triplet of bases on the mRNA is called a codon. The enzyme RNA polymerase joins the bases of the RNA to produce a complete single-stranded molecule. The base pairing is given in Table 4.2.

The molecule of mRNA detaches from the DNA template and the DNA strands can rejoin to remake the double helix. The mRNA leaves the nucleus via the nuclear pores and enters the cytoplasm.

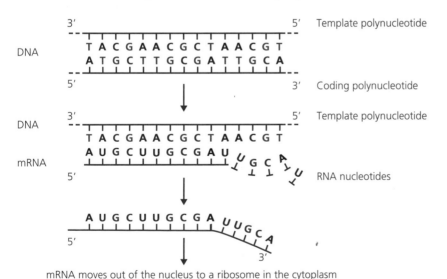

mRNA moves out of the nucleus to a ribosome in the cytoplasm

Figure 4.6 Transcription is the creation of a molecule of mRNA

Now test yourself

2 Explain why the nucleotides are read in triplets.

Answer on p. 117   TESTED ☐

Transcription is the conversion of the genetic code to a sequence of nucleotides in mRNA. A codon is a triplet of bases in the mRNA.

Translation is the conversion of the code in mRNA to a sequence of amino acids.

Table 4.2 Base pairing of the DNA template and mRNA

DNA template	mRNA
A	U
T	A
C	G
G	C

Now test yourself

3 Explain why the genetic code must be converted to mRNA before it leaves the nucleus.

Answer on p. 117   TESTED ☐

Translation

Translation is the conversion of the genetic code to a sequence of amino acids (Figure 4.7). The mRNA joins on to a ribosome in the cytoplasm (ribosomes consist of ribosomal RNA associated with enzymes). The ribosome contains enough space for two codons at a time. The mRNA slides through the groove between the two components of the ribosome. As each codon enters the ribosome, it is used to position the next amino acid. Amino acids must be activated — they are combined with a specific molecule of transfer RNA (tRNA) that has a specific base triplet called the anticodon. The anticodon of the tRNA is complementary to a codon on the mRNA. ATP is used in this activation process. The amino acids attached to tRNA molecules are aligned with the correct part of the mRNA by the complementary pairing of the bases in the codon and anticodon. Enzymes bind the amino acids together in a chain by condensation reactions to create a growing polypeptide chain. The tRNA molecule is then released to be reused. When the ribosome reaches the end of the mRNA, the complete polypeptide chain is released and folds to form the secondary and tertiary structure of the protein. Some proteins need to be activated by cyclic AMP (cAMP), which interacts with the new protein to alter its three-dimensional shape. This makes the protein a better shape to fit their complementary molecules.

Revision activity

Draw a diagram of a cell including the organelles relevant to protein synthesis, then draw a flow diagram of protein synthesis underneath.

Exam tip

Ensure that you understand how the structure of DNA and the different forms of RNA enable them to perform their functions.

Revision activity

Write a list of all the key terms used in this chapter, then add the meaning of each key term.

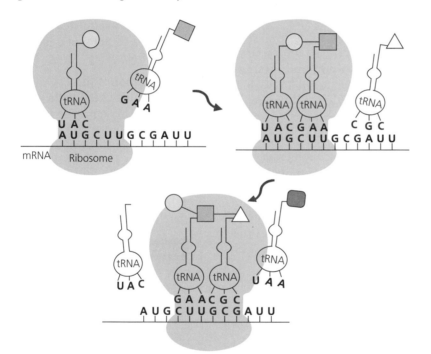

Figure 4.7 Translation: amino acids are aligned according to the triplet sequence in the mRNA

Now test yourself

4 Explain why mRNA must be single-stranded.

Answer on p. 117

TESTED

Exam practice

1 (a) Describe the structure of a DNA nucleotide. [3]
  (b) State two differences between the nucleotides of DNA and those of RNA. [2]
  (c) Explain why RNA molecules are much smaller than DNA molecules. [2]
2 In a well-known experiment, scientists allowed bacteria to grow on a medium that contained only heavy nitrogen. The bacteria incorporated this nitrogen into their DNA. After many generations, the scientists assumed that all the nitrogen in the DNA had been replaced by heavy nitrogen. They then placed the bacteria on a medium containing normal nitrogen and allowed the bacteria to breed. The DNA was extracted from the bacteria after successive generations and separated according to its mass. This separation technique involved centrifuging the DNA in a dense solution, which caused

Exam practice answers and quick quizzes at **www.hoddereducation.co.uk/myrevisionnotes**

bands of DNA to appear in the centrifuge tube. Each band represents DNA of a different mass. Heavier DNA appears in a band lower down the tube. The results are shown in the diagram below.

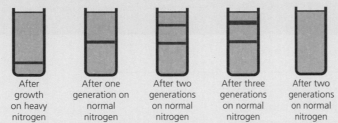

Banding pattern observed:

After growth on heavy nitrogen | After one generation on normal nitrogen | After two generations on normal nitrogen | After three generations on normal nitrogen | After two generations on normal nitrogen

(a) Which part of the DNA contains nitrogen? [1]

(b) The scientists believed that this experiment demonstrated that DNA is replicated in a semi-conservative fashion. Explain what is meant by *semi-conservative replication*. [3]

(c) Explain how the results in the diagram demonstrate semi-conservative replication. [4]

(d) If replication had been conservative — where whole molecules of DNA are conserved and totally new ones are made — the results would show a different banding pattern. On the far right-hand diagram above, show what banding you would expect to see using DNA collected from bacteria after two generations on normal nitrogen. [2]

3 The following table shows information about base ratios in three species.

Species	Percentage composition of bases			
	A	C	G	T
X	28.2	21.0	20.8	29.2
Y	24.3	24.9		
Z	34.2	15.8		

(a) State the full names of the bases A, C, G and T. [4]

(b) Complete the table. [2]

(c) Suggest why the total for species X is not exactly 100%. [2]

Answers and quick quiz 4 online

ONLINE

Summary

By the end of this chapter you should be able to:
- Describe the structure of nucleotides as monomers.
- Describe the structure of RNA and how it differs from that of DNA.
- Describe the structure of DNA as a double-stranded polynucleotide with four bases called adenine, thymine, cytosine and guanine.
- Describe how the bases pair in a complementary fashion.
- Describe ADP and ATP as phosphorylated nucleotides.
- Outline the semi-conservative method of DNA replication.
- Describe the genetic code.
- Describe transcription and translation as stages in the synthesis of polypeptides.

5 Enzymes

The role of enzymes

As intracellular and extracellular catalysts

REVISED

Enzymes are **globular proteins**. They act as **catalysts** to metabolic reactions in living organisms, which means they usually speed up metabolic reactions so that they occur at a reasonably fast pace even at body temperature.

Enzymes are required to build all the structures of the body (e.g. the cytoskeleton of a cell can be built up and reduced by enzyme activity), as well as to control the activity of the body.

Enzymes may be **intracellular** (working inside cells), such as **catalase** which converts hydrogen peroxide to oxygen and water. Alternatively, enzymes may be **extracellular** (working outside cells), such as the digestive enzymes **amylase** and **trypsin**, which are released into the digestive system.

A **globular protein** is a protein that folds and coils into a globular shape rather than a fibre.

Intracellular proteins work inside cells.

Extracellular proteins work outside cells.

The mechanism of enzyme action

Enzyme properties

REVISED

Enzymes have particular properties. These include:
- the molecule has a three-dimensional shape — its **tertiary structure**
- part of the molecule is an **active site** that is complementary to the shape of the substrate molecule
- each enzyme is specific to the substrate
- there is a high turnover number
- they have the ability to reduce the energy required for a reaction to occur
- their activity is affected by temperature, pH, enzyme concentration and substrate concentration
- the enzyme is left unchanged at the end of the reaction

Typical mistake

Candidates tend to describe enzymes 'breaking down' the substrate. It is far more precise to say 'hydrolyse' or 'oxidise', i.e. to name the actual reaction that occurs.

Exam tip

Recalling a list of the properties of enzymes is easy. You need to be able to *explain* how those properties are related to the structure of the enzyme molecules.

Specificity and the lock and key hypothesis

The **specificity** of an enzyme refers to its ability to catalyse just one reaction or type of reaction. Only one particular substrate molecule will fit into the active site of the enzyme molecule. This is because of the shape of the active site.

The shape of the active site is caused by the specific sequence of amino acids. This produces a specific tertiary structure — the three-dimensional shape of the molecule. This is referred to as the **lock and key hypothesis** (Figure 5.1).

The **lock and key hypothesis** explains how enzymes are specific to their substrate.

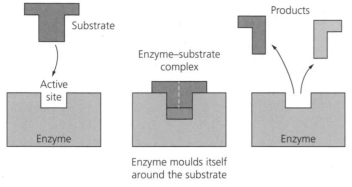

Figure 5.1 The enzyme has an active site that is complementary in shape to the substrate

Typical mistake

Candidates tend to refer to the lock and key hypothesis incorrectly. They make statements such as 'the enzyme works by the lock and key method', which is incorrect. The lock and key hypothesis explains how enzymes are specific to their substrate; it does not explain how they work.

Catalysing the reaction

Enzymes can speed up the rate of a reaction at body temperature. They lower the **activation energy** required for the reaction to occur. The activation energy is the amount of energy required to set off the reaction and break the bonds in the substrate molecule.

The induced-fit hypothesis

The **induced-fit hypothesis** helps to explain how the activation energy may be reduced.

The active site of an enzyme molecule does not have a perfectly complementary fit to the shape of the substrate. When the substrate moves into the active site, it interacts with the active site and interferes with the bonds that hold the shape of the active site. As a result, the shape of the active site is altered to give a perfect fit to the shape of the substrate. This changes the shape of the active site, which also affects the bonds in the substrate, making them easier to make or break (and therefore reducing the activation energy).

Typical mistake

Some candidates think that the lock and key hypothesis and the induced-fit hypothesis are mutually exclusive. However, the induced-fit hypothesis is a way to explain how a substrate moves into an active site that then changes to fit the substrate like a lock and key.

The course of an enzyme-controlled reaction

In an enzyme-controlled reaction, the enzyme (E) and substrate (S) molecules combine to form the **enzyme–substrate complex (ESC)**. The substrate is converted to the product, forming an **enzyme–product complex (ESP)**. The product is finally released and the enzyme is then free to take up another substrate molecule. This process is shown in Figure 5.2.

Figure 5.2 An enzyme-controlled reaction

Now test yourself

1 Using the analogy of a boulder in a hollow at the top of a hill, explain the role of an enzyme helping to overcome the activation energy in a reaction.

Answer on p. 117 [TESTED]

The **induced-fit hypothesis** is a hypothesis that modifies the lock and key hypothesis.

Now test yourself

2 Describe what bonds and interactions could cause the enzyme to change shape to wrap around the substrate more closely.

Answer on p. 117 [TESTED]

Typical mistake

Many candidates forget to mention the enzyme–substrate complex.

Effects of conditions on enzymes

pH

All enzymes have an optimum **pH** — the pH at which they work best. Therefore, they will not work as quickly at a pH outside their optimum range (Figure 5.3). This is because the hydrogen ions that cause acidity affect the interactions between *R* groups.

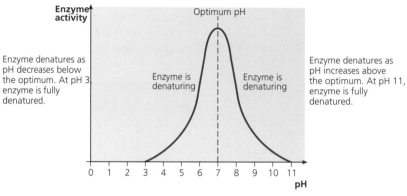

Figure 5.3 The effect of changing pH on enzyme activity

Altering the interactions between the *R* groups affects the tertiary structure of the molecule and may alter the shape of the active site. The shape will no longer be complementary to the shape of the substrate molecule.

> **Revision activity**
>
> Draw a graph to show the effect that pH change has on the activity of an enzyme. Annotate the graph with explanations for what happens at each stage.

Temperature

The effects of **temperature change** on enzyme action vary depending on the temperature range considered (Figure 5.4). Each enzyme has an optimum temperature at which it is most active. This temperature is often 37°C (in mammals), but it may be different in other organisms.

At low temperatures (0–45°C), the activity of most enzymes increases as temperature rises. At low temperatures, the molecules have little kinetic energy. They collide infrequently with the substrate molecules and activity is reduced. As temperature rises, the molecules gain more kinetic energy. They collide more frequently with the substrate molecules and are more likely to have sufficient energy to overcome the required activation energy. Therefore, activity increases.

At higher temperatures, enzymes lose their shape (they become denatured). Higher temperatures cause increased vibration of parts of the molecule. If the temperature rises above a certain point, the bonds within the enzyme molecule vibrate too much and break, which alters the bonding in the active site, changing its shape. The active site no longer fits the shape of the substrate and activity reduces quickly to zero.

> **Typical mistake**
>
> Many candidates seem to think that all enzymes have the same optimum temperature and pH, which is not the case.

> **Revision activity**
>
> Draw a graph to show the effect that temperature change has on the activity of an enzyme. Annotate the graph with explanations for what happens at each stage.

> **Typical mistake**
>
> When describing or explaining the effects of changing conditions on enzyme action, many candidates make statements such as 'temperature affects enzyme activity'. They forget to say that *changing* the temperature affects enzyme activity.

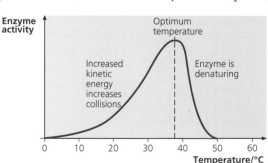

Figure 5.4 The effect of changing temperature on enzyme activity

Exam practice answers and quick quizzes at **www.hoddereducation.co.uk/myrevisionnotes**

Exam tip

To achieve full marks in the examination, you need to explain the effects of pH and temperature on the bonding in the active site and how this changes the shape of the active site. Remember that the changes are not instant — small increases above optimum will cause small changes and reduce activity. The enzyme does not suddenly stop working completely.

Enzyme concentration

REVISED

If there are more enzyme molecules in a particular volume of reaction medium, there are more active sites available. There is a greater likelihood of collisions between the enzyme and the substrate molecules. More interactions per second mean a higher rate of reaction. As the **enzyme concentration** increases, so does the rate of reaction (Figure 5.5).

The effect of pH and temperature on rates is usually because at extremes of pH and temperature some enzyme molecules are denatured and the concentration of active enzyme molecules is reduced.

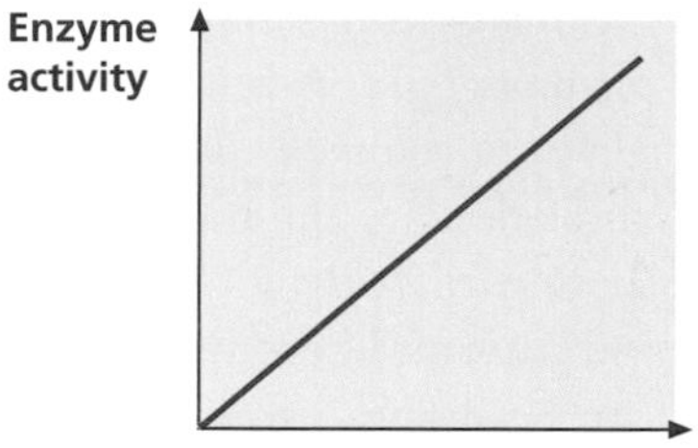

Figure 5.5 The effect of increasing enzyme concentration on enzyme activity

Substrate concentration

REVISED

If the **substrate concentration** is high, there is a greater chance of collisions between the enzyme and substrate molecules. As the substrate concentration increases, so does the rate of reaction (Figure 5.6).

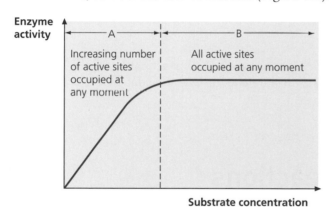

Figure 5.6 The effect of increasing substrate concentration on enzyme activity

If the number of enzyme molecules is limited, the rate of reaction levels off once all the enzyme active sites are fully occupied.

Exam tip

Questions could be limited to just one property, in which case it is likely that a lot of detail will be expected. Alternatively, this topic could be tested with a more open-ended type of question, in which case less detail will be expected on each property.

Revision activity

Draw a mind map to link together all the factors that affect enzyme activity and the explanations for each.

Practical investigations

Effects of pH, temperature, enzyme concentration and substrate concentration

REVISED

Exam tip

The written examination will include a number of marks to test your knowledge of practical work. These are often set in the context of a practical test carried out by a student or scientist. You may need to be able to describe a method, select suitable apparatus, evaluate a method, interpret results, analyse results or evaluate results.

You should be familiar with how the effects of changes in the following factors on enzyme activity can be investigated experimentally:

- pH
- temperature
- enzyme concentration
- substrate concentration

It is important to consider the following points.

1 *Volume and concentration of enzyme solution.*
2 *Volume and concentration of substrate solution.*
3 *Control of temperature.* A thermostatic water bath is often the best way.
4 *Control of pH.* A buffer solution controls pH.
5 *How can you make the test more reliable?* Repeat the test a number of times and calculate the mean.
6 *Testing reliability.* Comparing raw data to the mean is a good indication. There should be little variation around the mean — this can be shown using range bars. Calculating the standard deviation is even better.
7 *What is the control?* This is a test that omits one factor in the experiment to show that it is essential for the reaction to occur. It is usual to omit the enzyme from the reaction mixture to show that no reaction occurs without the enzyme.
8 *What level of precision is appropriate in the measurements?*
9 *How valid is the experiment?* Is it actually measuring what you think it should? Have you taken account of all possible conditions that may affect the reaction rate?

> **Exam tip**
>
> To gain full marks when describing a method, you must refer to the concentration and the volume of solutions.

> **Typical mistake**
>
> Many candidates refer to the amount of a substance when describing investigations. Ask yourself, what is the unit of 'amount'?

> **Revision activity**
>
> Draw the apparatus needed to collect and measure the volume of gas released from decomposing hydrogen peroxide. Annotate the diagram with information about how each variable can be controlled.

Now test yourself

TESTED ☐

3 Explain why it is important that only one variable is changed in a practical test.
4 Explain the difference between controlling a variable and the use of a control in a practical test.

Answers on p. 117

Enzyme-controlled reactions

Coenzymes and cofactors

REVISED ☐

Coenzymes are larger organic substances that take part in the reaction. They usually transfer other reactants between enzymes. Examples of coenzymes include coenzyme A, which takes part in aerobic respiration, and NAD, which is involved in transporting hydrogen atoms to the inner mitochondrial membrane. Coenzyme A and NAD are both made from the B vitamins.

Cofactors are inorganic substances, usually metal ions. They fit into the active site and activate the enzyme. Examples of cofactors include Cl^- in **amylase** and Zn^{2+} as a **prosthetic group** in **carbonic anhydrase**.

> **Coenzymes** are large organic substances that are involved in some chains of enzyme-controlled reactions.
>
> **Cofactors** are substances (usually a small metal ion) that are required to make enzymes function correctly.

Inhibitors

REVISED ☐

Inhibitors are substances that reduce the rate of reaction and fit into a site on the enzyme.

> **Inhibitors** are substances that reduce the activity of an enzyme.

Exam practice answers and quick quizzes at www.hoddereducation.co.uk/myrevisionnotes

Competitive inhibitors

Competitive inhibitors have a shape similar to the shape of the substrate. They fit into the active site, stopping the substrate molecules fitting in. This reduces the number of available active sites. The amount of inhibition depends on the relative concentrations of inhibitor and substrate molecules.

Non-competitive inhibitors

Non-competitive inhibitors fit into a different site on the enzyme molecule. They cause a change in the shape of the enzyme molecule. This affects the active site, so the substrate molecule can no longer fit in.

Reversible and non-reversible inhibitors

Reversible inhibitors occupy the enzyme site only briefly whereas **non-reversible inhibitors** bind permanently to the enzyme.

Drugs and poisons

Many **metabolic poisons** act by inhibiting enzymes. A poison such as cyanide inhibits the action of the enzyme cytochrome oxidase in aerobic respiration. Cytochrome oxidase contains iron ions and the cyanide binds to them. Many **medicinal drugs** also act as inhibitors of enzymes in the body. Aspirin binds to enzymes, preventing the formation of cell-signalling molecules that normally stimulate pain sensitivity. This is why taking the correct dose of medicinal drugs is important as overdosing can be lethal, especially if the inhibitor is non-reversible.

Product inhibition

Product inhibition occurs when the product of an enzyme-controlled reaction inhibits the enzyme. This can act to prevent too much product being formed.

Now test yourself

5 Explain how a non-competitive inhibitor prevents the substrate entering the active site.

Answer on p. 117

TESTED

Typical mistake

Many candidates think that competitive inhibitors are reversible and non-competitive inhibitors are irreversible, but this is not the case.

Revision activity

Write a list of all the key terms used in this chapter, then add the meaning of each key term.

Exam practice

1 Amylase is an enzyme that converts starch (amylose) to reducing sugars. A student investigated the effect of different temperatures on the rate of action of amylase. She prepared water baths containing starch suspension at four different temperatures. She then added amylase to each sample of starch and stirred. Each minute, the student transferred two drops of the starch/amylase mixture to a cavity tile containing iodine solution and noted the colour produced. The results are shown in the following table.

Time (min)	Temperature (°C)			
	5	20	40	70
1	Blue/black	Blue/black	Yellow	Blue/black
2	Blue/black	Blue/black	Yellow	Blue/black
3	Blue/black	Blue/black	Yellow	Blue/black
4	Blue/black	Dark yellow	Yellow	Blue/black
5	Blue/black	Yellow	Yellow	Blue/black
6	Blue/black	Yellow	Yellow	Blue/black
7	Blue/black	Yellow	Yellow	Blue/black

(a) (i)   What does the colour blue/black indicate? [1]
   (ii) What does the yellow colour in the samples at 20°C and 40°C indicate? [1]
(b) Suggest why the tube at 40°C was yellow after 1 minute. [2]
(c) Explain why the sample at 5°C remained blue/black. [3]
(d) Explain why the sample at 70°C remained blue/black. [4]
(e) As part of her evaluation, the student commented that it was difficult to sample all the tubes at the same time. Suggest one improvement she could make to her experiment. [1]
(f) She concluded that the optimum temperature for amylase activity is 40°C. Explain why this figure may not be accurate and suggest further improvements to her procedure that may make the test more accurate. [3]

2 (a) Enzymes are biological catalysts. Explain what is meant by the term *biological catalyst*. [2]
  (b) Explain why more than one enzyme is needed to digest starch. In your response, ensure that the properties of enzymes are linked to their structure. [7]

3 (a) A student performed a practical procedure in which the rate of formation of maltose was measured in the presence and absence of chloride ions. In the presence of chloride ions, the rate of maltose formation increased.
    (i) State the name given to a metal ion that increases the rate of an enzyme-controlled reaction. [1]
    (ii) Suggest how the chloride ions act to have this effect on the rate of reaction. [2]
  (b) The student extended his investigation to test the effect of temperature on the rate of reaction. When explaining his results, he made the following statement:

As the <u>heat</u> increased, the reaction went faster until it got to its <u>highest</u>. After this, the rate of reaction fell. This happened because the enzyme was <u>killed</u> and the hydrogen peroxide could not fit into the enzyme's <u>key</u> site.

Suggest a more appropriate word to replace each of the underlined words. [4]

Answers and quick quiz 5 online

ONLINE

Summary

By the end of this chapter you should be able to:
- State that enzymes are globular proteins with a specific tertiary structure, which catalyse metabolic reactions.
- State that enzyme action may be intracellular or extracellular.
- Describe the mechanism of action of enzyme molecules, with reference to specificity, active site, lock and key hypothesis, induced-fit hypothesis, enzyme–substrate complex, enzyme–product complex and lowering of activation energy.

- Describe and explain the effects of changes in pH, temperature, enzyme concentration and substrate concentration on enzyme activity.
- Explain the effects of competitive and non-competitive inhibitors on the rate of enzyme-controlled reactions, with reference to both reversible and non-reversible inhibitors.
- Explain the importance of cofactors and coenzymes in enzyme-controlled reactions.
- State that metabolic poisons and medicinal drugs may be enzyme inhibitors.

6 Biological membranes

Roles of membranes in cells

The plasma (cell-surface) membrane

The plasma membrane is a **partially permeable barrier** between the cell and its environment. It keeps the contents of the cell separate from its environment, and limits what molecules can enter and leave the cell. It acts as the site for certain chemical reactions and enables the cell to communicate with other cells through the process of **cell signalling**.

> **Exam tip**
>
> Always remember to refer to the outer membrane of a cell as a plasma membrane or cell-surface membrane.

> **Revision activity**
>
> Draw a diagram of a cell to show all the membranes inside the cell-surface membrane.

Internal membranes (organelle membranes)

Other membranes in the cell separate the **organelles** from the **cytoplasm**. These compartmentalise the cell, separating processes so that each process can occur in a specialised area of the cell. For example, all the enzymes involved in one process can be kept together and other processes do not interfere. Concentration gradients can be formed across the membranes. The membranes may act as the sites of specific **chemical reactions**, such as oxidative phosphorylation in aerobic respiration.

> **Revision activity**
>
> From memory, write a list of four functions of cell membranes.

> **Now test yourself**
>
>
> 1 Explain how a concentration gradient can be built up.
> 2 Suggest why a compartmentalised cell is more efficient than one that is not compartmentalised.
>
> Answers on p. 117

The fluid mosaic model of membrane structure

The **fluid mosaic model** describes the molecular arrangement of the membranes in a cell (Figure 6.1). A fluid mosaic membrane consists of:

- a bilayer of phospholipid molecules
- cholesterol which regulates the fluidity of the membrane, making it more stable
- glycolipids and glycoproteins that function in cell signalling or cell attachment
- protein molecules that float in the phospholipid bilayer. Some proteins are partially held on the surface of the membrane — these are called extrinsic proteins. Others are embedded in the membrane — these are called intrinsic proteins. Some proteins float freely in the bilayer whereas others may be bound to other components in the membrane or to structures inside the cell

> The **fluid mosaic model** is the accepted structure of membranes in a cell.

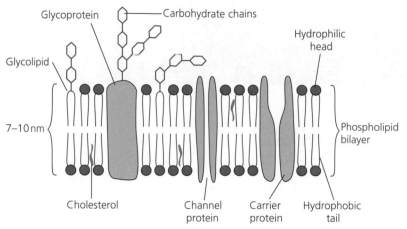

Figure 6.1 The fluid mosaic model of membrane structure

Phospholipids

Phosopholipids form a barrier that limits movement of some substances into and out of the cell, or into and out of the organelles, so the membrane is partially permeable. Small, fat-soluble substances dissolve into the phospholipid bilayer and diffuse across the membrane. Water-soluble molecules and ions cannot easily dissolve and cross the membrane. Small molecules like water may diffuse across slowly, but most require special transport mechanisms.

Cholesterol

Cholesterol fits between the tails of the phospholipid molecules. It inhibits movement of the phospholipids, reducing the fluidity of the membrane. It also holds the phospholipid tails together, for mechanical stability. Cholesterol makes the membrane less permeable to water and ions.

Gycolipids and glycoproteins

The carbohydrate group on the protein or lipid molecule always has a specific shape and is used to recognise the cell — to identify it as 'self' or 'foreign'. Antigens on cell surfaces are usually **glycolipids** or **glycoproteins**.

Drugs and **hormones** can bind to these **membrane-bound receptors**. Medicines can be made to fit the receptors on certain cells. For example, asthmatics take salbutamol, which fits the receptors on smooth muscle in the airways to cause relaxation.

Cells communicate in an organism by cell signalling to coordinate the activities of the organism. The shape of the glycoprotein or glycolipid may be complementary to the shape of a signalling molecule in the body. Such complementary shapes can be used as binding sites to which the signalling molecules (e.g. hormones and neurotransmitter molecules in a synapse) attach. If the correct binding site is not present, the cell cannot respond to the signalling molecule. Binding sites are also used for cell attachment — the cells of a tissue bind together to hold the tissue together.

Proteins

Proteins have a variety of functions, such as enzymatic activity and cell signalling. However, many functions involve moving substances across the membrane. For example, some proteins may form:

- pores that allow the movement of molecules that cannot dissolve in the phospholipid bilayer
- carrier molecules that allow facilitated diffusion
- active pumps

Revision activity

From memory, draw a small diagram of a part of a cell-surface membrane to show the variety of molecules involved in its structure.

Exam tip

Never describe the cholesterol as increasing rigidity of the membrane.

Revision activity

Draw a diagram of a membrane including all the molecules involved in its structure and annotate it to describe what each component does.

Exam tip

Always say that the shape of the signal molecule is complementary to the shape of the receptor molecule.

Revision activity

Draw a table listing each membrane component in the left-hand column and its function in the right-hand column.

Now test yourself

3 Explain how the membrane can be selectively permeable.

Answer on p. 117

TESTED ☐

Membrane structure and permeability

Temperature

Membranes are partially permeable, fluid and stable at normal body **temperature**. If temperature increases, the molecules gain kinetic energy and move about more. This increases the permeability of the membranes to certain molecules. Any molecules that diffuse through the phospholipid bilayer will diffuse more quickly. This is because as the phospholipids move about, they leave temporary gaps between them, providing space for small molecules to enter the membrane.

If temperature increases further, the phospholipid bilayer may lose its mechanical stability (it may melt) and the membrane becomes even more permeable. Eventually, the proteins in the membrane will denature. This will further damage the structure of the membrane and it will become completely permeable.

> **Typical mistake**
>
> Many candidates describe the membrane as denaturing, but it is only the proteins that denature.

> **Revision activity**
>
> Describe how a cell-surface membrane may be different from the membranes inside a cell.

Solvents

Solvents such as alcohol dissolve fatty substances. As the concentration of alcohol increases, the membrane is more likely to dissolve.

> **Now test yourself**
>
> 4 Explain the importance of complementary shapes in cell signalling.
>
> Answer on p. 117
>
> TESTED

The movement of molecules across membranes

Passive transport

Passive transport is the movement of molecules that does not need metabolic energy in the form of **adenosine triphosphate (ATP)**. It uses energy in the form of the kinetic (movement) energy. It only occurs when molecules move down a concentration gradient.

Because molecules move randomly, some may move in the 'wrong' direction — so you should describe passive transport as the net movement of molecules down their concentration gradient. Passive transport can occur in three forms:

- **Diffusion** (Figure 6.2) — the net movement of molecules away from a concentrated source. This may occur across a membrane if the molecules are fat-soluble or if they are small and can fit between the phospholipids in a membrane.
- **Facilitated diffusion** (Figure 6.2) — diffusion across a membrane that is helped by a **transport protein** in the membrane. The protein could be a pore protein (which may be gated) or it could be a carrier protein.
- **Osmosis** — the net movement of water molecules across a **partially permeable membrane**. Water molecules move down their **water potential gradient** (i.e. from an area of higher water potential to an area of lower water potential).

> **Passive transport** is the movement of molecules without the use of metabolic energy.
>
> **Diffusion** is the net movement of molecules down a concentration gradient.
>
> **Facilitated diffusion** is diffusion that is aided by a protein in the membrane.
>
> **Transport proteins** are proteins that help move substances across membranes. They include carrier proteins which may move molecules by changing shape.
>
> **Osmosis** is the movement of water from a region of higher water potential to a region of lower water potential.

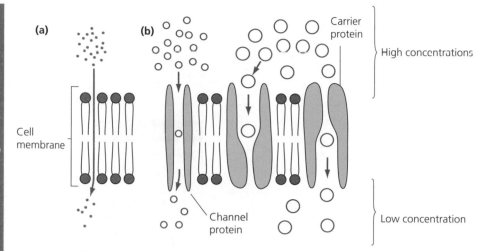

Figure 6.2 (a) Diffusion and (b) facilitated diffusion

The rate of diffusion

Diffusion occurs without using metabolic energy. It relies on the kinetic energy of the molecules. The rate of diffusion is affected by:

- Temperature — a higher temperature gives molecules more kinetic energy. At higher temperatures the molecules move faster, so the rate of diffusion increases.
- Concentration gradient — more molecules on one side of a membrane (or less on the other) increases the concentration gradient. This increases the rate of diffusion.
- Size of molecule — small molecules or ions can move more quickly than larger ones. Therefore, they diffuse more quickly than larger ones.
- Thickness of membrane — a thick barrier creates a longer pathway for diffusion, so diffusion is slowed down by a thick membrane.
- Surface area — diffusion across membranes occurs more rapidly if there is a greater surface area.

Active transport

Active transport (Figure 6.3) involves the movement of molecules using metabolic energy in the form of ATP. It can move molecules against their concentration gradient and uses membrane-bound proteins that change shape to move the molecules across the membrane.

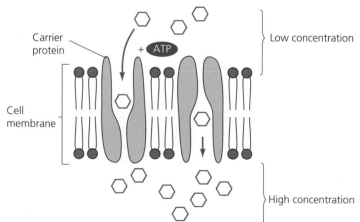

Figure 6.3 Active transport

> **Exam tip**
>
> Don't describe or explain osmosis in terms of 'water concentration' as this can be confused with the concentration of solutes. Always use the term *water potential*.

> **Now test yourself**
>
> 5 Explain why charged ions must be transported by facilitated diffusion rather than by simple diffusion.
>
> Answer on p. 117　　TESTED ☐
>
> REVISED ☐

> **Active transport** is the movement of molecules against a concentration gradient using metabolic energy in the form of ATP.

Bulk transport

Bulk transport is the movement of molecules through a membrane by the action of vesicles. **Endocytosis** is the formation of vesicles by the plasma membrane, which moves molecules into the cell. **Exocytosis** is the fusion of vesicles with the plasma membrane, which moves molecules out of the cell. Bulk transport uses metabolic energy.

> **Revision activity**
>
> Draw a mind map to show how a range of different substances pass through cell membranes.

Now test yourself

6 Explain why proteins pass through membranes by bulk transport.

Answer on p. 117

Water potential and osmosis

Water potential

Pure water has a **water potential** of zero. As solutes (sugars or salts) are added to a solution, the water potential gets lower. Therefore, a salt solution has a water potential below zero, i.e. a negative potential.

Water molecules will move from a solution with a higher water potential to a solution with a lower (more negative) water potential. Therefore, water molecules always move down their water potential gradient.

> **Water potential** is a measure of the concentration of free water molecules.

> **Typical mistake**
>
> Some candidates confuse water potential and water potential gradient. A gradient can only exist between two places.

Osmosis

A cell placed in water has a lower (more negative) water potential than the surrounding water. There is a water potential gradient from high outside the cell to lower inside the cell. As a result, water molecules enter the cell.

A cell placed in a strong salt solution has higher (less negative) water potential than the surrounding solution. There is a water potential gradient from higher inside the cell to lower outside the cell, so water molecules leave the cell.

> **Exam tip**
>
> It may be easier to describe osmosis in terms of water molecules moving from a less negative region to a more negative region.

Table 6.1 The effects of osmosis on animal and plant cells

Cell type	Solution	
	Pure water	**Strong salt**
Animal	An animal cell has no cell wall. The plasma membrane has no strength, so the animal cell will burst as water enters the cell.	An animal cell has no cell wall. The cytoplasm will shrink and the cell will shrivel. Its appearance is known as crenelated.

| Cell type | Solution | |
	Pure water	Strong salt
Plant	Water makes a plant cell turgid. The vacuole is full of watery sap and the cytoplasm pushes the plasma membrane out against the cell wall. The cell wall is strong and will stop the cell bursting.	A plant cell will lose its turgidity. It will become flaccid. If the water loss continues, the cell vacuole will shrink. The cytoplasm will also shrink and the plasma membrane pulls away from the cell wall. This is called plasmolysis.

Now test yourself

TESTED

7 Explain why a plant cell will not burst when placed in pure water.
8 Using the terms *water potential* and *water potential gradient*, explain why a plant cell loses turgidity when placed in a strong salt solution.
9 In a plasmolysed plant cell, state what is found in the gap between the cell wall and the plasma membrane. Explain how it gets there.

Answers on pp. 117–18

Exam practice

1 (a) Describe how small molecules such as water and carbon dioxide can cross a plasma membrane. [2]
  (b) Describe how large molecules can be brought into a cell. [3]
2 (a) The diagram below shows four animal cells that touch each other. Each cell has a different water potential as shown by the figures. Draw arrows to show the direction in which water will move by osmosis from cell to cell. [4]

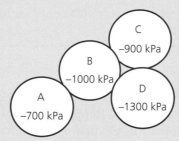

  (b) Using the term *water potential*, explain the movement of water that would occur if cell C in the diagram were placed in pure water. [3]
3 Describe the functions of the following components of a cell membrane:
  (a) cholesterol [1]   (b) a glycoprotein [2]   (c) phospholipids [2]
4 Use the idea of cell signalling to describe how the cell ensures that vesicles containing proteins can be transported to the correct organelle inside a cell. [4]

Answers and quick quiz 6 online

ONLINE

Summary

By the end of this chapter you should be able to:
● Outline the roles of membranes in cells.
● Describe the fluid mosaic model of membrane structure.
● Describe the roles of membrane components.

● Outline the effect of factors such as temperature and solvents on membrane permeability and structure.
● Understand how substances pass through membranes.
● Understand the terms *water potential* and *water potential gradient*.

7 Cell division, cell diversity and cellular organisation

The cell cycle

Processes during the cell cycle

REVISED

The cell cycle (Figure 7.1) is a series of events during which the cell duplicates its contents and splits in two. **Mitosis** is a small part of the cycle. The remainder of the cycle (known as **interphase**) is used for copying the **chromosomes** and checking the genetic information. During interphase, the cell also increases in size, produces new organelles and stores energy for another division.

> **Mitosis** is the division of cells to produce two genetically identical daughter cells.

	Stage of cell cycle	Events within the cell	Control of cycle — checkpoints
INTERPHASE	G₁ phase (growth)	Cells grow and increase in size Transcription of genes (mRNA made) Protein synthesis occurs Organelles duplicate	G_1 cyclin-CDK complexes promote the production of transcription factors needed to produce enzymes for DNA replication G_1 checkpoint ensures that the cell is ready for DNA synthesis
	S phase (synthesis)	DNA replicates, producing pairs of identical sister chromatids	Active S cyclin-CDK complexes ensure all DNA is replicated once
	G₂ phase (growth)	Cells grow	G_2 checkpoint ensures that the cell is ready to enter M phase
	M phase (nuclear division or mitosis)	Cell growth stops Nuclear division consisting of stages: prophase, metaphase, anaphase and telophase Cytokinesis (cytoplasmic division)	Mitotic cyclin-CDK complexes promote the production of the spindle and the condensation of chromosomes Metaphase checkpoint ensures that the cell is ready to complete mitosis

Figure 7.1 The cell cycle

Now test yourself

TESTED

1 Explain why the cell must copy the DNA before mitosis.
2 Explain why the cell must produce new organelles and store energy during interphase.

Answers on p. 118

Mitosis

The significance of mitosis in life cycles

REVISED

Mitosis produces two genetically identical cells which are used for:
1 **growth** of the organism
2 **repair of tissues**

3 replacement of old cells
4 asexual reproduction

Typical mistake

Some candidates lose marks because they suggest that mitosis is for the growth or repair of cells. This is not true — cells grow during interphase and mitosis repairs tissues, not cells.

The main stages of mitosis

REVISED

Four stages make up mitosis in the cell cycle (Figure 7.2).

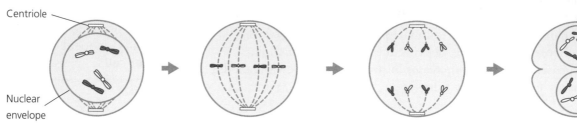

Centriole

Nuclear envelope

Prophase

- The start of mitosis
- Chromosomes shorten and thicken as DNA is tightly coiled
- Each chromosome is visible as two chromatids joined at the centromere
- Prophase ends as the nuclear envelope breaks into small pieces
- Centrioles organise fibrous proteins into the spindle

Metaphase

- Chromosomes are held on the spindle at the middle of the cell
- Each chromosome is attached to the spindle on either side of its centromere

Anaphase

- Chromatids break apart at the centromere and are moved to opposite ends of the cell by the spindle

Telophase

- Nuclear envelopes reform around the chromatids that have reached the two poles of the cell
- Each new nucleus has the same number of chromosomes as the original, parent cell
- The nuclei are genetically identical to each other

Figure 7.2 The stages of mitosis

Exam tip

Remember that the cell cycle and mitosis are continuous processes and the names of the stages reflect parts of a continuum.

Meiosis

The significance of meiosis in life cycles

REVISED

Meiosis is an alternative form of cell division. It produces four cells that are:

- not genetically identical
- gametes
- **haploid** (contain half the normal number of chromosomes)

The cells produced by meiosis are gametes — sex cells used for sexual reproduction. These cells contain one chromosome from each pair of **homologous chromosomes**. These have:

- the same shape and size
- the centromere in the same position
- the same genes in the same positions on the chromosomes

Typical mistake

Many candidates confuse meiosis with the process of fertilisation, which takes place after meiosis. In fertilisation, two gametes fuse to restore the diploid state in which the cell contains the homologous chromosomes.

Exam practice answers and quick quizzes at **www.hoddereducation.co.uk/myrevisionnotes**

The main stages of meiosis

Before division starts (during interphase), the DNA replicates so that each chromosome consists of two identical copies called sister **chromatids** which are held together by a centromere. The cell divides twice and each division can be divided into four stages (Figure 7.3).

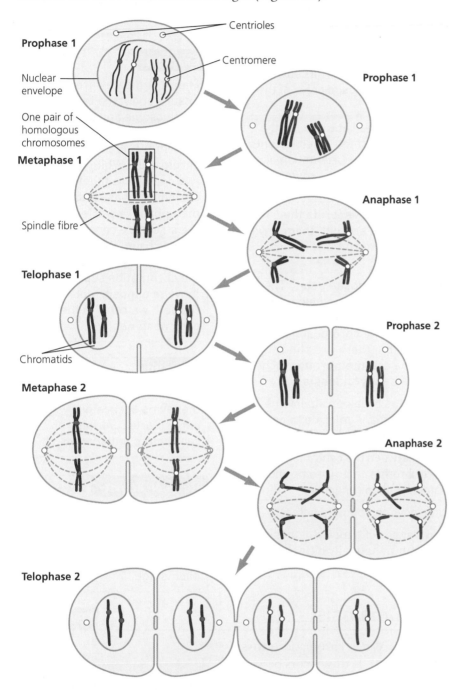

Figure 7.3 The stages of meiosis

Table 7.1 The stages of meiosis

Stage of meiosis	Events within the cell
Prophase 1	Chromosomes condense — they supercoil to become shorter and thicker. Homologous chromosomes pair to form **bivalents** containing four chromatids. The chromatids in each bivalent break and rejoin to form **chiasmata** or crossovers — this is where sections of the non-sister chromatids can be exchanged. The nuclear membrane breaks up to form small membrane sacs. The **centriole** replicates and migrates to opposite poles of the cell and forms the **spindle fibres**.
Metaphase 1	Microtubules attach from the centrioles to the centromere of each chromosome. The bivalents move to the equator of the cell. Orientation of each bivalent on the equator is random — maternal or paternal chromosomes could be facing either pole.
Anaphase 1	The microtubules shorten to separate the homologous chromosomes and pull them towards opposite poles. Each chromosome still consists of two chromatids.
Telophase 1	The chromosomes reach opposite poles. The nuclear membrane reforms around each set of chromosomes to produce two nuclei. These nuclei are haploid as they have one chromosome from each homologous pair (but there are still two sister chromatids). The **cell membrane** pinches in to form two cells — this is **cytokinesis**.
Prophase 2	The nuclear membranes break up again. The centrioles replicate again and migrate to opposite poles of the two new cells.
Metaphase 2	Microtubules attach between the centrioles and the centromere of each chromosome. The chromosomes move to the equator and align randomly.
Anaphase2	Sister chromatids move to opposite poles.
Telophase 2	The nuclear membranes reform. Cytokinesis occurs to produce four genetically different haploid cells.

A **bivalent** is the pair of homologous chromosomes.

Chiasmata are the points at which chromatids cross over (singular: chiasma).

Exam tip

Many candidates are unsure about the terms *chromosome* and *chromatid*. A chromosome replicates to form two identical chromatids. While the two copies are together during prophase and metaphase, we refer to the structure consisting of two chromatids as a chromosome. However, once the chromatids separate and move to opposite poles of the cell, we refer to them as chromosomes.

Exam tip

Make sure you can spell the terms accurately, particularly *centriole*, *centromere* and *meiosis*.

Revision activity

Draw your own set of diagrams showing meiosis and annotate each sketch to show what is occurring at each stage.

Creating genetic variation in meiosis

REVISED

Genetic variation can be created by meiosis. When chromatids **cross over**, they exchange lengths of DNA (Figure 7.4). If this occurs between non-sister chromatids, it makes new combinations of alleles.

Figure 7.4 Exchanging genetic material between non-sister chromatids

Now test yourself

3 Explain why there must be two divisions in meiosis.

Answer on p. 118

TESTED

Exam practice answers and quick quizzes at **www.hoddereducation.co.uk/myrevisionnotes**

The way the bivalents orientate on the equator during metaphase 1 is random. This means that either the maternal or the paternal chromosome of a bivalent my face either pole. Therefore, the combination of maternal and paternal chromosomes migrating to either pole is random. This is called **independent assortment** of homologous chromosomes (Figure 7.5). In a similar way, the orientation of the chromosomes on the equator in metaphase 2 is random. Therefore, the combination of chromatids migrating to each pole is random. This is called independent assortment of sister chromatids.

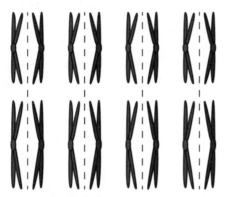

Figure 7.5 Independent orientation of homologous chromosomes on the equator creates genetic diversity by independent assortment

Cell specialisation in multicellular organisms

Cell specialisation for particular functions

REVISED

Cells of **multicellular organisms** are **specialised** in a number of ways, as shown in Table 7.2.

Squamous means flattened.

Table 7.2 Cell specialisation

Cell type and function	Specialisation	How that specialisation helps its function
Erythrocytes (red blood cells) carry oxygen in the blood	Small and flexible	To fit through tiny capillaries
	Full of haemoglobin	To bind to the oxygen
	No nucleus	To allow more space for haemoglobin
	Biconcave shape	To provide a large surface area to take up oxygen quickly
Neutrophils engulf and digest foreign matter or old cells	Flexible shape	To enable movement through tissues
	Lobed nucleus	To help movement through membranes
	Many ribosomes	To manufacture digestive enzymes
	Many lysosomes	To hold digestive enzymes
	Many mitochondria	To release the energy needed for activity
	Well-developed cytoskeleton	To enable movement
	Membrane-bound receptors	To recognise materials that need to be destroyed

Cell type and function	Specialisation	How that specialisation helps its function
Sperm carry the paternal chromosomes to the egg	Tail (flagellum)	To enable rapid movement
	Acrosome	To help digest egg surface
	Small	To make movement easier
	Many mitochondria	To release the energy needed for rapid movement
Epithelial cells act as surfaces	Often flat (**squamous**)	To cover a large area
	Often thin (squamous)	To provide a short diffusion distance
	May be ciliated	To move mucus
	May be cuboid	To provide a barrier
	Many glycolipids and glycoproteins in cell-surface membrane	To hold cells together or for cell signalling
Palisade cells for photosynthesis	Elongate	To fit many chloroplasts into the space
	Contain many chloroplasts	To absorb as much light as possible
	Show cytoplasmic streaming	To move the chloroplasts around
	Contain starch grains	To store products of photosynthesis
Root hair cells absorb water and mineral ions from the soil	Long extension (hair)	To increase surface area
	Active pumps in cell-surface membrane	To absorb mineral ions by active transport
	Thin cell wall	To reduce barrier to movement of ions and water
Guard cells control the stomatal opening	Active pumps in cell-surface membrane	To move mineral ions in and out of cell to alter the water potential
	Unevenly thickened wall	To cause the cell to change shape as it becomes more turgid
	Large vacuole	To take up water and expand to open the stoma

Revision activity

Draw a detailed diagram of each cell type in Table 7.2 and annotate it to explain how the cell is specialised.

Now test yourself

4 Explain how cells differentiate.

Answer on p. 118

TESTED

Tissues, organs and organ systems

Tissues

REVISED

A **tissue** is a collection of cells that work together to perform a particular function. They may be similar to each other or they may perform slightly different roles. For example:

1 **Squamous epithelium** is a layer of flattened cells bound together to produce a surface.
   - **Ciliated epithelium** contains ciliated cells that move mucus over their surface and goblet cells that produce the mucus.

2 Cartilage consists of cells called chondrocytes that secrete a matrix of collagen.

3 Muscle is found in three types: smooth muscle consists of single cells that can contract; skeletal muscle forms multinucleate fibres containing protein filaments that slide passed one another; cardiac muscle forms cross-bridges to ensure that the muscle contracts in a squeezing action.

- **Xylem** contains vessels that carry water and xylem fibres for support.
- **Phloem** contains two types of cell — sieve tube elements which form sieve tubes, and companion cells.

Now test yourself

5 Explain why forming tissues is more efficient than using individual cells to perform a task.

Answer on p. 118

TESTED

Organs

REVISED

An **organ** is a collection of tissues working together to perform a common function.

Organ systems

REVISED

An **organ system** is made up of two or more organs working together to perform a life function such as excretion or transport.

Revision activity

Write separate lists of all the tissues and organs in the human transport and gaseous exchange systems.

The features and differentiation of stem cells

Stem cells and differentiation

REVISED

Stem cells are cells that are not specialised or differentiated. They maintain the capacity to undergo mitosis and differentiate into a range of cell types. Differentiation is the ability of a cell to specialise to form a particular type of cell.

Stem cells are a **renewing source** of **undifferentiated cells** for the growth and repair of tissues and organs. During growth and repair, stem cells divide to produce new cells, which then differentiate to become specialised to their function. Stem cells have the ability to use all their genes. Differentiation occurs by switching on or off appropriate genes.

The production of blood cells

REVISED

Stem cells in the bone marrow divide and differentiate to form both red and white blood cells.

Erythrocytes (red blood cells)

Erythrocytes are specialised to carry oxygen. See Table 7.2 for specialisations. They have no nucleus and very few organelles, providing more space for haemoglobin molecules which are synthesised during development before the other organelles are lost.

Neutrophils (white blood cells)

Neutrophils are the most common type of phagocyte used to ingest and destroy bacteria. See Table 7.2 for specialisations.

The production of xylem vessels and phloem sieve tubes

Xylem and phloem are transport tissues in plants. New cells are produced by mitosis in the **meristem**. These cells are expanded by the uptake of water and the development of a vacuole. They then differentiate into xylem and phloem.

Xylem

Lignin is deposited in their cell walls to strengthen and waterproof the wall. The cells die and the contents are removed as the end walls break down, forming continuous columns of cells. These form tubes with wide lumens to carry water and dissolved minerals. The lignification is incomplete in some places, forming bordered pits.

Phloem

Phloem consists of two types of cell that work together:
1 Sieve tube elements lose their nucleus and most of their organelles. The end walls develop numerous sieve pores to form sieve plates between the elements.
2 Companion cells retain their organelles and can carry out metabolism to obtain and use ATP to actively load sugars into the sieve tubes.

Sieve tube elements and companion cells are linked by numerous **plasmodesmata**.

> **Plasmodesmata** (singular: plasmodesma) are connections between cells where the cytoplasm is continuous.

The potential uses of stem cells in research and medicine

Stem cells can be sourced from different tissues:
1 embryonic stem cells, which are present in a young embryo
2 blood from the umbilical cord
3 adult stem cells found in developed tissues such as **bone marrow** (but scientists are finding stem cells in almost all tissues)
4 scientists can also induce certain tissue cells to become stem cells (known as induced pluripotent stem cells or iPS cells).

Stem cells can be used in the following ways in research and medicine.

Repair of damaged tissues

Stem cells have been used to treat certain conditions and continued research suggests that many uses can be found:
1 Stem cells in bone marrow are used to treat diseases of the blood, such as leukaemia.
2 Stem cells have been used to repair the spinal cord of rats.
3 Stem cells have been used to treat mice with type 1 diabetes.
4 Stem cells in the retina can be made to produce new light-sensitive cells.
5 Stem cells directed to become nerve tissue could be used to treat **neurological conditions** such as **Alzheimer's disease** and **Parkinson's disease**.
6 Stem cells may also be used to treat other conditions such as arthritis, stroke, burns, blindness, deafness and heart disease.

Developmental biology

Scientists use stem cells to gain a better understanding of how multicellular organisms develop, grow and mature.

1 They study how differentiation occurs — how cells develop to make particular cell types.
2 They study what happens when differentiation goes wrong.
3 They are trying to find out if they can re-enable differentiation and growth in adult cells to help tissue repair (healing) in later life or even the ability to re-grow an organ or limb.

Revision activity

Write a list of all the key terms used in this chapter, then add the meaning of each key term.

Exam practice

1 The following statements are about meiosis.
   A Chromatids pair during prophase 1.
   B Bivalents form during prophase 2.
   C Homologous chromosomes are independently assorted.
   D Four haploid cells are produced.
   E Meiosis is used for growth and repair.
   Which of the following options identifies the correct statements? [1]
   (a) All five statements are correct.
   (b) Statements B, C and D are correct.
   (c) Statements C and D are correct.
   (d) All statements are incorrect.
2 (a) Explain what is meant by the term *differentiation*? [2]
   (b) Beta cells in the pancreas are specialised to produce the protein hormone insulin. Suggest how these cells may be specialised. [3]
   (c) Diabetes is a disease in which the beta cells stop producing insulin. Suggest how stem cells could be used to cure this disease. [2]
   (d) Explain the advantages of using stem cells from an embryo. [2]
3 (a) Define the terms *tissue* and *organ*. [4]
   (b) Plant transport tissues are called xylem and phloem. Describe how cells are organised to form these tissues. [5]

Answers and quick quiz 7 online

ONLINE

Summary

By the end of this chapter you should be able to:
● Describe the cell cycle and the stages of mitosis.
● Explain the significance of mitosis for growth, repair and asexual reproduction.
● Describe the stages of meiosis and explain the significance of meiosis in producing haploid cells.

● Understand the meanings of the terms *diploid, differentiation, haploid, homologous chromosomes, stem cell, tissue, organ* and *organ system*.
● Describe and explain how certain cells and tissues are specialised for their function.
● Explain how stem cells can be used in research and medicine.

8 Exchange surfaces

The need for specialised exchange surfaces

Surface area to volume ratio

REVISED

A living organism needs to absorb substances from its surroundings and remove waste products. This can only occur through its surface area. However, as an organism increases in size, its volume increases so it needs more from its environment. Unfortunately, its surface area does not increase as quickly as its volume, so the larger an organism gets, the more difficult it becomes to absorb enough substances over its surface. This can be demonstrated by considering a simple set of data:

- Assume that the organism is a cube with sides of length l.
- Its surface area is $6 \times$ the area of one side or $6 \times$ length squared $(6 \times l^2)$.
- Its volume is length $\times$ length $\times$ length or length cubed (l^3).

Table 8.1 shows what happens to surface area, volume and **surface area to volume ratio (SA:V)** as an organism increases in size.

Table 8.1 The effects on an organism as it increases in size

Length of organism (l) (mm)	Surface area of organism ($6 \times l^2$) (mm²)	Volume of organism (l^3) (mm³)	Surface area to volume ratio
1	6	1	6 : 1
5	150	125	1.2 : 1
10	600	1000	0.6 : 1

We can see that as size increases:
- surface area increases
- volume increases, but more quickly than surface area
- surface area to volume ratio decreases

The significance of surface area to volume ratio

Single-celled organisms are small and have a large surface area to volume ratio. Their surface area is large enough for sufficient oxygen and nutrients to diffuse into the cell to provide all its needs, and for wastes to diffuse out.

Multicellular organisms have a smaller surface area to volume ratio. Diffusion is too slow for the oxygen and nutrients to diffuse across the whole organism. The surface area is no longer large enough to supply all the needs of the larger volume. Therefore, a specialised exchange surface is required, such as the lungs which are used for gaseous exchange. Being multicellular and **metabolically active** also increases the need for a specialised exchange surface.

The **surface area to volume ratio (SA:V)** is the surface area of an organism divided by its volume. It is a key concept as the surface area must be able to provide sufficient oxygen through diffusion from the environment.

Revision activity

Draw a graph of surface area to volume ratio plotted against body size for a range of body lengths from 5 μm (bacteria) up to 5 m (small whale). Remember to convert the units appropriately. Mark the graph to show where an amoeba, a mouse, a man and an elephant would fit on the graph.

Exam tip

Remember that:
- length, surface area and volume all have different units
- surface area to volume ratio has no units
- suitable units must always be included in any work involving figures

Typical mistake

Many candidates confuse surface area with surface area to volume ratio. An elephant has a large surface area, but it has a small surface area to volume ratio.

Now test yourself

1 List the factors that affect the need for a specialised surface for gaseous exchange.
2 Explain why a single-celled organism such as an amoeba does not need a specialised surface for gaseous exchange whereas a large tree does.

Answers on p. 118

Efficient exchange surfaces

The features of good gaseous exchange surfaces

A good gaseous exchange surface (Figure 8.1) must be able to exchange gases quickly enough to provide for the activity of the cells inside the organism. The surface has certain features, as shown in Table 8.2.

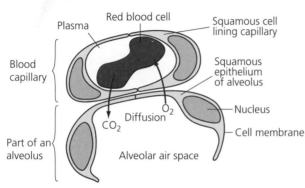

Figure 8.1 The gaseous exchange surface in the lungs

Table 8.2 The features of good surfaces for gaseous exchange

Feature	Reason	In the lungs
Large surface area	To provide space for molecules of oxygen and carbon dioxide to pass	Lung epithelium folded to form numerous alveoli (singular: alveolus)
Thin layer	To provide a short diffusion pathway	Lung epithelium and capillary endothelium are made from squamous cells
Steep concentration gradient	To ensure molecules diffuse rapidly in the correct direction	Good supply of blood on one side and ventilation of the air sacs on the other side

Steep concentration gradient

A steep **concentration gradient** is maintained by increasing the concentration of molecules on the supply side and reducing the concentration on the demand side. In the lungs, this is achieved by good blood flow and ventilating the air spaces. Blood flow brings carbon dioxide to the lungs and removes oxygen; ventilation brings oxygen to the lung surface and removes carbon dioxide. The presence of a thin barrier to diffusion helps to create a steep concentration gradient.

Now test yourself

3 List the factors that affect the concentration gradient.

Answer on p. 118

> **Typical mistake**
>
> Many candidates describe the lungs as having a 'thin cell wall' — they probably mean a 'wall of thin cells' or a 'thin wall of cells'. This sort of vague wording should be avoided. Describe the barrier as creating a short diffusion pathway because the cells are squamous.

> **Revision activity**
>
> Draw a mind map to link the features of a good gaseous exchange surface to rate of diffusion.

> A **concentration gradient** is the difference in concentration between two points.

> **Exam tip**
>
> Remember to describe changes in the concentration of the gases in the blood or in the air sacs as it is the concentration gradient that drives diffusion.

Ventilation and gaseous exchange in mammals

The lungs

Figure 8.2 shows the structure of the human gaseous exchange system. Table 8.3 summarises the distribution and function of lung cells and tissues.

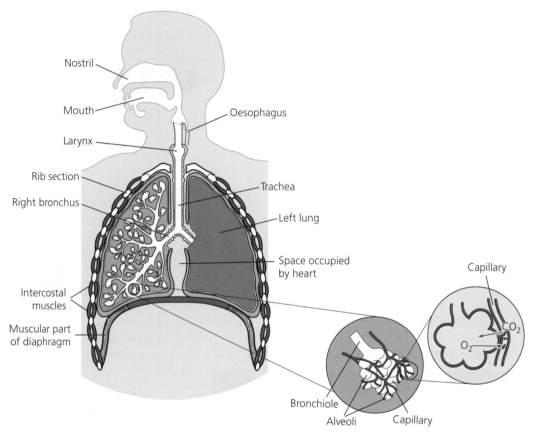

Figure 8.2 The human gaseous exchange system with details of the alveoli

Table 8.3 The distribution and function of cells and tissues in the lungs

Structure	Distribution	Function
Capillaries	Over surface of alveoli	To provide a large surface area for exchange
Cartilage	In walls of bronchi and trachea	To hold the airways open
Ciliated epithelium	On surface of airways	The cilia move or waft the mucus along
Elastic fibres	In walls of airways and over alveoli	To recoil to return the airway or alveolus to original shape. In alveolus this helps to expel air
Goblet cells	In ciliated epithelium	To produce and release mucus
Smooth muscle	In walls of airways	Contracts to constrict or narrow the airways
Squamous endothelium	Capillary wall	To provide a thin barrier to exchange — a short diffusion pathway
Squamous epithelium	Surface of alveoli	To provide a thin barrier to exchange — a short diffusion pathway

Figures 8.3 and 8.4 show details of the wall of the trachea and the distribution of tissues in the lungs.

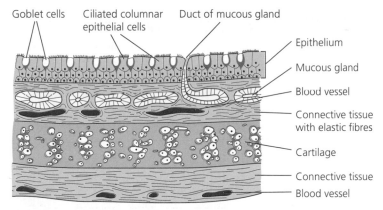

Figure 8.3 The wall of the trachea

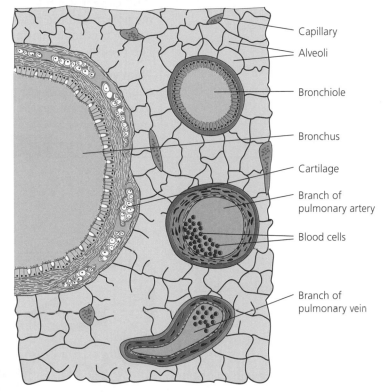

Figure 8.4 The distribution of tissues in the lungs

Ventilation

REVISED

Ventilation is also known as breathing. It refreshes the air in the alveoli. Ventilation is achieved by the action of the diaphragm and the intercostal muscles. The processes that take place during inspiration and expiration are summarised in Table 8.4.

Ventilation means breathing and refreshing the air in the alveoli.

Table 8.4 Inspiration and expiration

Structure/feature	Inspiration (inhaling)	Expiration (exhaling)
Diaphragm	Contracts and moves downwards, pushing organs down	Relaxes and is pushed up by the organs underneath
Intercostal muscles	Contract to raise the rib cage up and outwards	Relax and allow the rib cage to fall
Volume change	Chest cavity increases in volume	Chest cavity reduces in volume

Structure/feature	Inspiration (inhaling)	Expiration (exhaling)
Pressure change	Pressure inside chest cavity reduces and falls below atmospheric pressure	Pressure inside chest cavity rises above atmospheric pressure
Air movement	Air is pushed into lungs by higher atmospheric pressure	Air is pushed out of lungs by higher pressure in alveoli

> **Exam tip**
>
> To achieve full marks, you will need to describe all the volume and pressure changes accurately.

Vital capacity and tidal volume

REVISED

Vital capacity is the maximum volume of air that can be breathed in or out when taking a deep breath. Vital capacity is typically 4.5 dm³ in young men and 3.0 dm³ in young women. Vital capacity can be increased through training. Singers and athletes often have a large vital capacity.

Tidal volume is the volume of air breathed in and then out in one breath. The tidal volume changes according to the needs of the body. At rest it is usually about 0.5 dm³.

> **Vital capacity** is the maximum volume of air that can be breathed in or out in one breath.
>
> **Tidal volume** is the volume of air breathed per breath, usually taken at rest.

Breathing rate and oxygen uptake

REVISED

Using a spirometer

A **spirometer** is an apparatus for measuring the volume of air inspired and expired by the lungs (Figure 8.5).
1 The subject should wear a nose clip to ensure that no oxygen escapes from the system and no additional air is added.
2 The subject breathes through the mouthpiece.
3 As the subject inhales, oxygen is drawn from the air chamber, which therefore descends.
4 As the subject exhales, the air chamber rises again.
5 Air returning to the air chamber passes through the canister of soda lime, which absorbs carbon dioxide.
6 The movements of the air chamber are recorded by a data logger or on a revolving drum.
7 Tidal volume is measured simply by allowing the subject to breathe normally.
8 Vital capacity is measured by asking the subject to breathe out as deeply as possible.

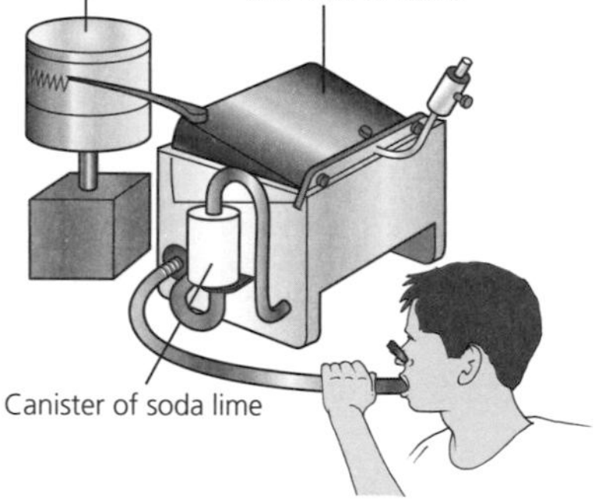

Revolving drum   Oxygen chamber floating in a tank of water

Canister of soda lime

Figure 8.5 A spirometer

Analysing the trace

All measurements are taken from the spirometer trace (Figure 8.6). Always remember to measure at least three readings (if possible) and calculate a mean. **Breathing rate** is calculated by counting the number of peaks in 1 minute. **Oxygen uptake** is a little more difficult:
● As carbon dioxide is removed, the total volume in the air chamber decreases.
● The volume of oxygen absorbed is shown by the difference in height of the last peak from the first peak during normal breathing.
● Divide this volume by time taken in order to calculate the rate of oxygen uptake.

> **Now test yourself**
>
> 4 Explain why the subject should wear a nose clip.
> 5 Explain the function of the soda lime and why it is essential.
>
> **Answers on p. 118**
>
> TESTED

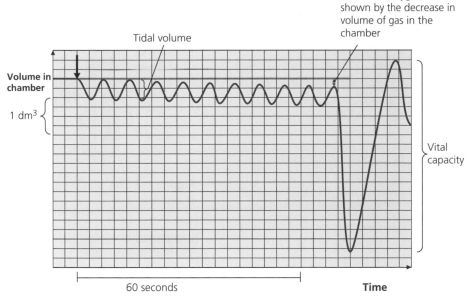

Tidal volume

Volume of oxygen used is shown by the decrease in volume of gas in the chamber

Volume in chamber

1 dm³

Vital capacity

60 seconds

Time

Figure 8.6 Measurements can be taken from a spirometer trace

Ventilation and gaseous exchange in bony fish and insects

Bony fish

REVISED

Fish must exchange gases with the water in which they live (Figure 8.7). They use gills to absorb oxygen dissolved in the water and release carbon dioxide into the water. Each gill consists of two rows of **gill filaments** (primary lamellae) attached to a bony arch. The filaments are very thin and their surface is folded into many **gill lamellae (gill plates)**. This provides a large surface area. Blood capillaries carry deoxygenated blood close to the surface of the secondary lamellae where exchange takes place. The blood flows in the opposite direction to the flow of water (a **countercurrent flow**). Ventilation is achieved by movements of the floor of the mouth (**buccal cavity**) and **operculum** (the bony flap over the gills).

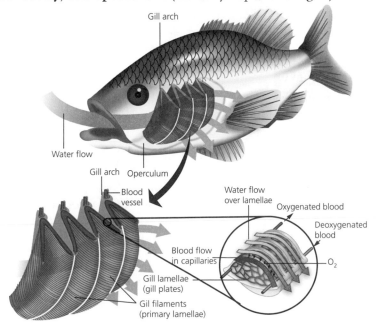

Gill arch

Water flow

Gill arch Operculum

Blood vessel

Water flow over lamellae

Oxygenated blood

Deoxygenated blood

Blood flow in capillaries

O₂

Gill lamellae (gill plates)

Gil filaments (primary lamellae)

Figure 8.7 The gills in a bony fish

Insects

Insects do not transport oxygen in blood. They have an air-filled tracheal system that supplies air directly to all the respiring tissues (Figure 8.8). Air enters the system via pores called **spiracles**. The air passes through the body in a series of tubes called tracheae (singular: **trachea**). These divide into smaller tubes called tracheoles. The ends of the tracheoles open into **tracheal fluid**. Gaseous exchange occurs between the air in the tracheole and the tracheal fluid.

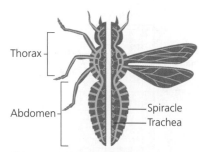

Figure 8.8 The tracheal system of an insect

Larger insects can also ventilate their tracheal system by movements of the body. This can be achieved by:

1 flexible air sacs that are squeezed by contraction of the leg or wing muscles

2 changes in the volume of the thorax caused by contraction of the wing muscles (**thoracic movement**)

3 changes in the volume of the abdomen caused by specialised muscles (**abdominal movement**). This is often coordinated with opening and closing the spiracles, inhaling at the front end and exhaling at the rear end

> **Revision activity**
>
> Write a list of all the key terms used in this chapter, then add the meaning of each key term.

Exam practice

1 Which row in the following table correctly matches the function to the tissue? [1]

Row	Elastic tissue	Smooth muscle	Squamous epithelium	Ciliated epithelium
A	Reduce diameter of the lumen	Recoil to return alveoli to original size	Provide a short diffusion distance	Trap and remove bacteria
B	Recoil to return alveoli to original size	Reduce diameter of the lumen	Provide a short diffusion distance	Trap and remove bacteria
C	Contract to return alveoli to original size	Reduce diameter of the lumen	Provide a short diffusion distance	Trap and remove bacteria
D	Recoil to return alveoli to original size	Reduce diameter of the lumen	Trap and remove bacteria	Provide a short diffusion distance

2 (a) Explain why a large, active animal such as a mammal needs a specialised surface for gaseous exchange. [3]

(b) The following table shows how the surface area and volume of a sphere changes as its size increases.

Diameter (cm)	Surface area (cm²)	Volume (cm³)	Surface area to volume ratio
2	50.3	33.5	1.5 : 1
5	314.2	523.7	0.6 : 1
10	1256.8	4189.3	

(i) Calculate the surface area to volume ratio of a sphere of 10 cm diameter. Show your working. [2]

(ii) Describe the trend shown by the surface area to volume ratio as the size of the sphere increases. [2]

3 (a) State one function in the airways of each of the tissues listed below.
elastic tissue   ciliated epithelium   smooth muscle [3]

(b) Describe how ciliated cells and goblet cells work together to reduce the risk of infection in the lungs. [3]

(c) The alveoli walls contain elastic fibres. Suggest what may happen to the alveoli if the elastic fibres are damaged. [2]

(d) In asthmatics, certain substances in the air cause the smooth muscle in the walls of the airways to contract. Suggest the effect this will have on the person. [2]

4 (a) Describe how you would use a spirometer to measure tidal volume. [3]

(b) Explain why the air chamber should be filled with medical-grade oxygen rather than air. [2]

(c) Describe two other precautions that should be taken when using a spirometer. [2]

Answers and quick quiz 8 online

ONLINE

Summary

By the end of this chapter you should be able to:

- Understand the importance of surface area to volume ratios.
- Describe the features of a good gaseous exchange surface.
- Describe the features of the lungs that make them a good surface for gaseous exchange.
- Describe the distribution of tissues in the lungs and explain the role of each in an efficient organ of gaseous exchange.

- Outline the mechanism of breathing.
- Explain the terms *vital capacity* and *tidal volume*.
- Understand how a spirometer can be used to measure vital capacity, tidal volume, breathing rate and oxygen uptake.
- Describe the gaseous exchange system in bony fish and insects.

9 Transport in animals

The need for transport systems in multicellular animals

Size, metabolic rate and surface area to volume ratio

REVISED

All living animal cells need a supply of oxygen and nutrients to survive. They also need to remove waste products such as carbon dioxide and urea so that they do not build up and become toxic. The three main factors that affect the need for a transport system are **size**, **metabolic rate** and **surface area to volume ratio**. You should recall that size and surface area to volume ratio have been explained in Chapter 8. Active organisms usually have a high metabolic rate. This requires more oxygen to allow more aerobic respiration to take place. This is essential so that more ATP can be released from food to provide energy for the higher level of activity. The oxygen and substrate molecules such as glucose must be supplied to the active cells rapidly, which is why active organisms require a transport system.

> **Typical mistake**
>
> Many candidates describe small organisms as having a large surface area rather than a large surface area to volume ratio, or they describe a large organism as having a small surface area.

> **Exam tip**
>
> Don't confuse surface area with surface area to volume ratio. It may help to always write SA:V rather than surface area to volume ratio.

Different types of circulatory systems

Single and double circulatory systems

REVISED

In a **single circulatory system**, blood flows through the heart once every time it goes around the body. Fish have a single circulatory system. The blood flows from the heart to the gills and then on to the body before returning to the heart:

heart → gills → body → heart

In a **double circulatory system**, blood flows through the heart twice for every circuit around the body. Mammals have developed a circulation that involves two separate circuits. One circuit carries blood to the lungs to take up oxygen. This is the pulmonary circulation. The other circuit carries the oxygen and nutrients around the body to the tissues. This is the systemic circulation. The heart is adapted to form two pumps, one for each circulation.

body → heart → lungs → heart → body

> **Typical mistake**
>
> Many candidates describe a single circulation as 'blood going through the heart once'. However, the blood must go around the body and return to the heart, so it is better to describe it as 'going through the heart once for every circuit of the body'.

Now test yourself

TESTED

1 Explain why a double circulatory system is more efficient than a single circulatory system.

Answer on p. 118

Open and closed circulatory systems

Insects have an **open circulatory system**. In an open system:

1 there is no separate tissue fluid
2 blood circulates around the organs and cells
3 pressure cannot be raised to help circulation
4 circulation is affected by body movements
5 oxygenated and deoxygenated blood mix freely

Fish and **mammals** both have a **closed circulatory system**. In a closed system:

- blood is kept in vessels
- pressure can be maintained
- pressure can be higher
- flow can be faster
- flow can be directed to certain tissues or organs

> **Revision activity**
>
> Draw an insect and a mammal. Inside each illustration, include a simple version of the circulation system.

The structure and functions of blood vessels

Arteries, arterioles, capillaries, venules and veins

Blood flows through a series of vessels. Each is adapted to its particular role in relation to its distance from the heart. All types of blood vessel have an inner layer or lining made of cells called the **endothelium**. This is a thin layer that is particularly smooth to reduce friction with the flowing blood. Figure 9.1 shows cross-sections of an artery and a vein. Table 9.1 compares the structure and functions of **arteries**, **veins** and **capillaries**.

> The **endothelium** is a thin layer of cells that lines all blood vessels.

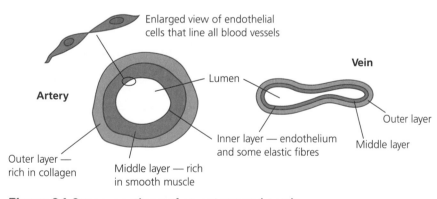

Figure 9.1 Cross-sections of an artery and a vein

Arterioles are small arteries with a spiral layer of smooth muscle. They distribute blood from the arteries to the capillaries and can be constricted to reduce blood flow. **Venules** are small veins that collect the blood from the capillaries and lead into the veins.

Table 9.1 Comparing the structure and functions of arteries, veins and capillaries

Feature	Arteries	Veins	Capillaries
Function	Transport the blood away from the heart	Transport the blood back to the heart	Enable the exchange of materials between the blood and tissue fluid
Thickness of wall	Thick	Thin	Very thin (one cell thick)
Components of wall	Endothelium lining surrounded by a thick middle layer of elastic tissue and smooth muscle, then a thick outer layer rich in collagen	Endothelium lining surrounded by a thinner layer of elastic tissue and smooth muscle, with a thin outer layer containing collagen	One layer of endothelium cells
Blood pressure	High	Low	Low
Presence of valves	No	Yes	No
Cause of flow	Pressure created by the heart, maintained by recoil of elastic tissues	Squeezing action of body muscles and valves to ensure correct direction	Pressure from action of the heart

Typical mistake

Many candidates describe the role of smooth muscle as 'creating a smooth surface to reduce friction', or they describe the action of the smooth muscle as 'pumping blood along the artery'. The smooth muscle is used to reduce the diameter of the vessel to maintain blood pressure or to reduce blood flow in a particular direction. Similarly, the elastic tissue is used to recoil the artery wall to maintain blood pressure.

Revision activity

Write a list the features of the arteries that maintain blood pressure and another list of the features that enable arteries to withstand high blood pressure.

Now test yourself

TESTED

2 Describe how the structure of an artery is adapted to its role of transporting blood at high pressure.
3 Describe how the structure of a vein is adapted to its role of transporting blood at low pressure.

Answers on p. 118

Blood, tissue fluid and lymph

Blood

REVISED

Blood is the fluid found inside the blood vessels. It consists of:
- water-based plasma containing dissolved substances — oxygen, nutrients such as glucose and amino acids, lipoproteins, carbon dioxide (most transported as bicarbonate ions), other wastes such as urea, hormones, antibodies and plasma proteins
- red blood cells (erythrocytes), probably carrying oxygen
- white blood cells (phagocytes), such as neutrophils and lymphocytes
- platelets

Tissue fluid

Tissue fluid surrounds the body cells. It is the plasma that has been filtered out of the blood, so it contains all the dissolved elements of the blood except the cells, platelets and plasma proteins. These are too large to pass out of the blood vessels. There may be some phagocytic neutrophils in tissue fluid as these can change shape to squeeze out of the blood vessels.

The formation of tissue fluid

The walls of the capillaries are a single layer of endothelium cells. Fluid and dissolved substances in the fluid can squeeze between the endothelium cells. The fluid is acted upon by two forces:
1 the **hydrostatic pressure** gradient between the blood and the tissue fluid, which tends to push fluid out of the capillary
2 the **oncotic pressure** gradient between the blood and tissue fluid, which tends to move fluid into the blood because the water potential of the blood is lower than water potential of the tissue fluid

At the arterial end of the capillary, the hydrostatic pressure created by the heart is still quite high. Therefore, there is a steep pressure gradient pushing fluid out of the capillary. This overcomes the oncotic pressure gradient and fluid moves out of the capillary to become tissue fluid. Oxygen and nutrients move into the tissues with the fluid.

At the venous end of the capillary, the hydrostatic pressure is lower. The hydrostatic pressure gradient is less steep than the oncotic pressure gradient and fluid returns to the capillary. Carbon dioxide and other wastes are carried back into the blood as tissue fluid moves back into the capillary. Some of the tissue fluid is drained into the blind-ending lymph vessels to become lymph.

> **Exam tip**
>
> Remember that tissue fluid is blood that does not contain blood cells or plasma proteins, but it does contain the dissolved components.

> **Hydrostatic pressure** is the pressure a fluid exerts on the sides of a vessel.
>
> **Oncotic pressure** is the osmotic pressure created by dissolved substances such as proteins.

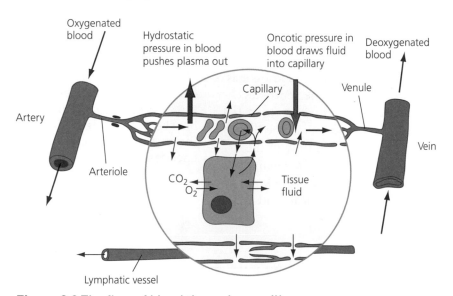

Figure 9.2 The flow of blood through a capillary

Lymph

Lymph is excess tissue fluid that is not returned to the blood vessel. Instead, it is drained into the lymph vessels. These carry the fluid back to the circulatory system by a different route.

Lymph contains the same substances as tissue fluid but it has less oxygen and glucose as these have been used by the cells. Lymphocytes produced in the lymph nodes may also be present.

The mammalian heart

The **mammalian heart** is a muscular pump that is divided into two sides. The right side pumps **deoxygenated blood** to the lungs to be oxygenated. The left side pumps **oxygenated blood** to the rest of the body. On both sides the action of the heart is to squeeze the blood, putting it under pressure and forcing it along the arteries.

External structure

REVISED

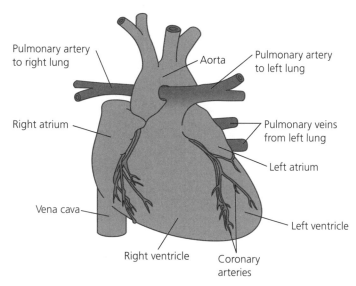

Figure 9.3 The external features of the heart

The muscle surrounding the two main pumping chambers (the ventricles) is dark red. Above the ventricles are two thin-walled chambers known as the atria (singular: atrium). These are much smaller than the ventricles. On the surface of the heart are coronary arteries. These carry oxygenated blood to the heart muscle itself. These arteries are important as the heart continually works hard. If they become constricted, blood flow to the heart muscle is restricted and this can reduce the delivery of oxygen and nutrients, such as fatty acids and glucose, causing a heart attack. At the top of the heart are the veins that carry blood into the heart and the arteries that carry blood out of the heart.

Internal structure

REVISED

The heart is divided into four chambers: two atria and two ventricles.

The atria

The two upper chambers are **atria**. These receive blood from the major veins. Deoxygenated blood flows from the vena cava into the right atrium. Oxygenated blood flows from the pulmonary vein into the left atrium. The atria have very thin walls as they do not need to create much pressure. Blood simply flows through the atria into the ventricles. When the ventricles are nearly full, the atrial walls contract just to completely fill the ventricles.

> **Revision activity**
>
> Draw a table to compare the compositions of blood, tissue fluid and lymph.

> **Deoxygenated blood** transports less oxygen and more carbon dioxide than oxygenated blood.
>
> **Oxygenated blood** transports oxygen to the organs.

> **Revision activity**
>
> Draw a diagram to show the external features of the heart and annotate it with the names and functions of each feature.

> The **atria** are the small chambers at the top of the heart, which collect blood from the main veins.

Exam practice answers and quick quizzes at **www.hoddereducation.co.uk/myrevisionnotes**

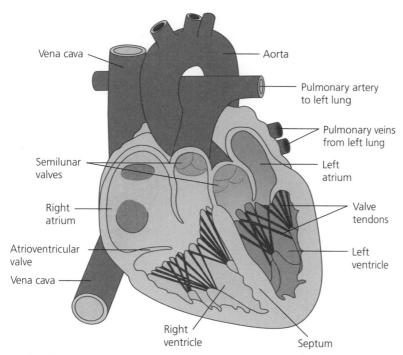

Vena cava
Aorta
Pulmonary artery to left lung
Pulmonary veins from left lung
Semilunar valves
Left atrium
Right atrium
Valve tendons
Atrioventricular valve
Left ventricle
Vena cava
Right ventricle
Septum

Figure 9.4 The internal features of the heart

The ventricles

The two lower chambers are the **ventricles**. Each has a thick muscular wall. The wall contracts to create pressure which pushes the blood into the arteries.

The right ventricle has walls that are thicker than the atrial walls. This enables it to pump blood out of the heart. The right ventricle pumps deoxygenated blood to the lungs. Therefore, the blood does not need to travel far. Also, the lungs contain many fine capillaries. The pressure of the blood must not be too high to prevent the capillaries in the lungs bursting.

The walls of the left ventricle are often two or three times thicker than those of the right ventricle. The blood from the left ventricle is pumped out through the aorta and needs sufficient pressure to propel it all the way to the extremities of the body. The pressure created by the left ventricle is typically measured at 110–120 mmHg.

The ventricles are separated from each other by a wall of muscle called the **septum**. This ensures that the oxygenated blood in the left side of the heart and deoxygenated blood in the right side are kept separate.

> The **ventricles** are the larger lower chambers, which have thick walls to pump the blood out of the heart.
>
> The **septum** is the muscle that separates the two ventricles.

Revision activity

Draw a diagram to show the internal features of the heart and annotate it with the names and functions of each feature.

Now test yourself

TESTED

4 Explain why the left ventricle looks so much bigger than the right.

Answer on p. 118

Major arteries

Deoxygenated blood leaving the right ventricle flows into the pulmonary artery leading to the lungs. Oxygenated blood leaving the left ventricle flows into the aorta. This carries blood to a number of arteries that supply all parts of the body.

The cardiac cycle

It is important that the chambers of the heart all contract in a coordinated fashion. If the chambers were to contract out of sequence, this would lead to inefficient pumping. The sequence of events involved in one heart beat is called the **cardiac cycle** (Figure 9.5).

> The **cardiac cycle** is the series of events in one heart beat.

Figure 9.5 The cardiac cycle

1 Blood returns to the heart from the body (via the vena cava) and lungs (via the pulmonary vein). Both the atria fill at the same time. The **valves** between the atria and the ventricles (the **atrioventricular valves**) are open to allow blood to flow straight through into the ventricles.

2 Once the ventricles are nearly full, the **sinoatrial node (SAN)** initiates a new heartbeat. It creates a wave of excitation which spreads over the walls of the atria. The walls contract, pushing a little extra blood from the atria into the ventricles. The wave of excitation is stopped by a layer of non-conducting fibres between the atria and the ventricles. The wave of excitation can only pass through the **atrioventricular node (AVN)**, where it is delayed a little. This allows time for the ventricles to fill.

3 After the delay, the wave of excitation passes down the bundle of His in the septum between the ventricles. At the base of the septum, the bundle splits into separate fibres called **Purkyne tissue** that carry the excitation up the walls of the ventricles, causing contraction from the base (apex) upwards. The walls of the two ventricles contract together. As the pressure rises, the atrioventricular valves are pushed shut which prevents blood re-entering the atria. The tendinous cords attached to the valves prevent them from inverting.

4 The blood pressure in the ventricles rises quickly until it rises above the pressure in the aorta and pulmonary artery. This pushes the **semilunar valves** open and blood is pushed into the main arteries.

5 Once contraction is complete, the muscles relax and the elasticity of the walls causes recoil to return the ventricles to their original size and shape. This causes the pressure in the ventricles to drop quickly. When the pressure drops below the pressure in the main arteries, the semilunar valves are pushed shut, preventing re-entry of blood into the ventricles. When the pressure in the ventricles drops below the pressure in the atria, the atrioventricular valves are pushed open by the pressure in the atria. This allows blood to flow into the ventricles again.

> The **atrioventricular valves** lie between the atria and the ventricles.
>
> The **semilunar valves** lie at the entrance to the main arteries.

> **Revision activity**
>
> Draw an outline of the heart and add arrows to show the direction of blood flow.

Now test yourself

5 Explain how the atrioventricular valves are opened and closed.
6 Explain why the electrical stimulation wave must be delayed at the AVN.

Answers on p. 118

Pressure changes during contraction

The changes in pressure in the heart can be represented by a graph (Figure 9.6). The important points are where one line crosses another — this is where the pressure in one chamber rises above that in another chamber, causing a valve to open or close.

> **Exam tip**
>
> Remember that it is the walls of the atria and ventricles that contract.

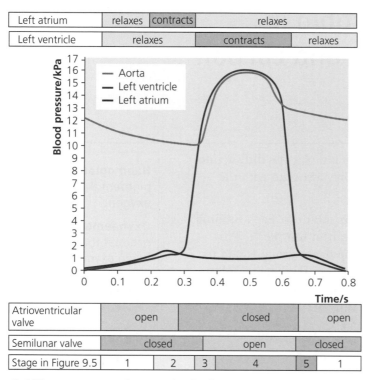

Figure 9.6 The pressure changes in the heart

Now test yourself

7 Describe and explain what happens at the point in Figure 9.6 where the line for the pressure in the left ventricle rises above the line for the pressure in the left atrium.

Answer on p. 119

Electrocardiograms

An **electrocardiogram (ECG)** (Figure 9.7) records the electrical activity of the heart. Wave P is the excitation of the atria. Wave QRS is the excitation of the ventricles. Wave T is associated with ensuring the muscles have time to rest. Abnormal heart activity can often be identified by an abnormal ECG trace. The waves may be smaller, inverted or further apart.

> **Exam tip**
>
> Questions are likely to show you a normal trace and an abnormal trace and ask you identify the differences.

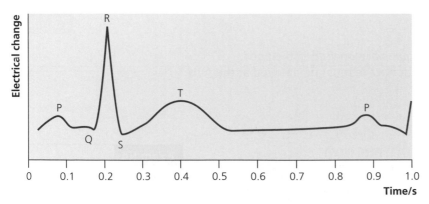

Figure 9.7 A normal ECG trace

The role of haemoglobin in transporting oxygen and carbon dioxide

Transport of oxygen

Oxygen enters the blood in the lungs. Oxygen molecules diffuse into the blood plasma and red blood cells. Here, they associate with the **haemoglobin** (Hb) to form **oxyhaemoglobin**.

Haemoglobin is a complex protein with four subunits. Each subunit contains a haem group that contains a single iron ion (Fe^{2+}). This attracts and holds one oxygen molecule. The haem group is said to have an **affinity** for (an attraction to) oxygen — the haemoglobin attracts and holds the oxygen. Each haem group can hold one oxygen molecule, so each haemoglobin molecule can carry four oxygen molecules.

> **Haemoglobin** is the red pigment that transports oxygen.
>
> **Oxyhaemoglobin** is the product that is formed when oxygen from the lungs combines with haemoglobin in the blood.

Oxygen dissociation curves

Haemoglobin has a high affinity for oxygen. The amount it takes up depends on the amount of oxygen in the surrounding tissues, which is measured by its partial pressure of oxygen (pO_2) or oxygen tension.

At a low pO_2, haemoglobin does not readily take up oxygen molecules. The haem groups are hidden at the centre of the haemoglobin molecule, making it difficult for the oxygen molecules to associate with them. This difficulty in combining the first oxygen molecule accounts for the low saturation level of haemoglobin at low pO_2.

As the pO_2 rises, one oxygen molecule succeeds in associating with one of the haem groups. This causes a conformational change in the shape of the haemoglobin molecule, which allows more oxygen molecules to associate with the other three haem groups more easily. This accounts for the steepness of the curve as pO_2 rises (Figure 9.8). In the body tissues, cells need oxygen for aerobic respiration, so oxyhaemoglobin releases the oxygen. This is called **dissociation** and it is shown in an **oxygen dissociation curve**.

> **Dissociation** is the release of oxygen from oxyhaemoglobin.

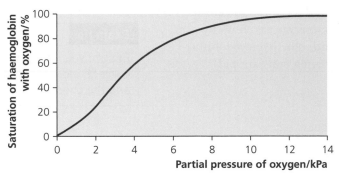

Figure 9.8 An adult oxygen dissociation curve

Now test yourself

TESTED

8 Explain why a person using a spirometer containing air would soon feel tired and breathless.

Answer on p. 119

Fetal haemoglobin

The haemoglobin of a mammalian fetus has a higher affinity for oxygen than adult haemoglobin. Its haemoglobin must be able to take up oxygen from an environment that makes the adult haemoglobin release oxygen. In the placenta, the **fetal haemoglobin** must absorb oxygen. This reduces the oxygen tension near the blood, making the maternal haemoglobin release oxygen. Therefore, the oxyhaemoglobin dissociation curve for fetal haemoglobin is to the left of the curve for adult haemoglobin.

> **Fetal haemoglobin** is a modified form of haemoglobin found in the mammalian fetus.

Transport of carbon dioxide

REVISED

Carbon dioxide released from respiring tissues must be removed from the tissues and transported to the lungs. Carbon dioxide in the blood is transported in three ways:
- as hydrogencarbonate ions (HCO_3^-) in the plasma (85%)
- combined directly with haemoglobin to form a compound called carbaminohaemoglobin (10%)
- dissolved directly in the plasma (5%)

How are hydrogencarbonate ions formed?

As carbon dioxide diffuses into the blood, some of it diffuses into the red blood cells. Here, it combines with water to form a weak acid (carbonic acid). This reaction is catalysed by the enzyme **carbonic anhydrase**.

$$CO_2 + H_2O \rightarrow H_2CO_3$$

The carbonic acid dissociates to release hydrogen ions (H^+) and hydrogencarbonate ions (HCO_3^-).

$$H_2CO_3 \rightarrow H^+ + HCO_3^-$$

The hydrogencarbonate ions diffuse out of the red blood cells into the plasma. The charge inside the red blood cells is maintained by the movement of chloride ions (Cl^-) from the plasma into the red blood cells. This is called the **chloride shift**.

The hydrogen ions could cause the contents of the red blood cells to become acidic. To prevent this, the hydrogen ions are taken up by haemoglobin to produce **haemoglobinic acid**. The haemoglobin is acting as a buffer (a compound that can maintain a constant pH).

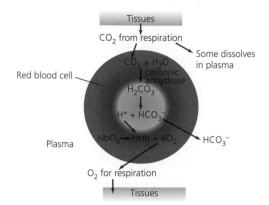

Figure 9.9 How carbon dioxide is converted to hydrogencarbonate ions

The Bohr effect

The hydrogen ions produced are absorbed by the haemoglobin. However, in very active tissues a lot of carbon dioxide is released and, therefore, a lot of hydrogen ions are created. As the concentration of hydrogen ions increases, this decreases the pH of the cytoplasm. The change in pH alters the structure of the haemoglobin, reducing its affinity for oxygen. Therefore, more oxygen is released. The haemoglobin dissociation curve shifts to the right (**the Bohr effect**). This also frees up more haemoglobin molecules to absorb the additional hydrogen ions.

The Bohr effect is the shift to the right of the position of the haemoglobin dissociation curve in the presence of extra carbon dioxide.

Revision activity

Write a list of all the key terms used in this chapter, then add the meaning of each key term.

Exam practice

1 (a) Describe the initiation and coordination of one heart beat. [5]
  (b) (i) In a condition known as superventricular tachycardia, the non-conducting fibres between the atria and the ventricles occasionally conduct the excitation wave. Suggest what effect this might have on the action of the heart. [2]
    (ii) Suggest what effects this might have on the patient. [2]
2 Explain why artery walls contain both smooth muscle and elastic fibres. [4]
3 (a) State what creates the hydrostatic pressure at the arterial end of the capillary. [2]
  (b) State two reasons why the pressure is much lower at the venous end of the capillary. [2]
4 Describe what is meant by an open circulatory system and explain why it is less efficient than a closed system. [3]

Answers and quick quiz 9 online

ONLINE

Summary

By the end of this chapter you should be able to:
- Explain the need for transport systems in large and active organisms.
- Explain the difference between open/closed circulatory systems and single/double circulatory systems.
- Describe the structure of the blood vessels.
- Understand the differences between blood, tissue fluid and lymph.
- Describe the external and internal structures of the mammalian heart.
- Describe the action of the heart and the cardiac cycle.
- Understand how oxygen and carbon dioxide are transported.

10 Transport in plants

The need for transport systems in multicellular plants

Size, metabolic rate and surface area to volume ratio

The need for a transport system in large and active organisms is explained in Chapter 9.

The structure and function of the vascular system

Roots, stem and leaves

Multicellular plants are large organisms so they need a transport system (Figure 10.1). Xylem tissue moves water and minerals from the roots to the leaves. Phloem tissue moves assimilates up and down the plant from sources to sinks.

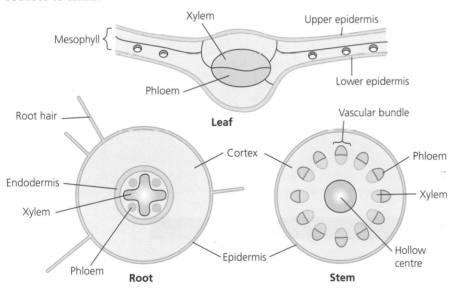

Figure 10.1 The distribution of transport tissues in a leaf, stem and root

> **Revision activity**
>
> From memory, draw the position of the xylem and phloem in roots and stems.

Xylem vessels

Xylem vessels are adapted to enable the free flow of water along the vessels. The cells have been killed by impregnation of the walls with lignin.

Adaptations of xylem

Xylem vessels have several adaptations (Figure 10.2):
- end walls are removed to form long tubes

- no cytoplasm or organelles are present
- cell walls are impregnated with lignin (lignified) to make the vessel wall waterproof and to strengthen the vessel to prevent it collapsing
- spiral, annular and reticulate thickening strengthens the wall to prevent collapse
- bordered pits between the vessels allow the movement of water between vessels

Exam tip

Make sure you say that the *cell walls* have been lignified — don't say that the vessels or the xylem have been lignified. Remember that this makes the walls waterproof — the xylem itself is not waterproof as there are pits to allow water in and out.

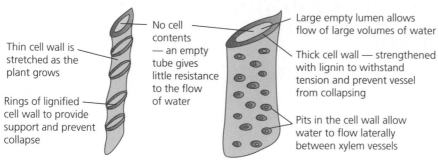

Thin cell wall is stretched as the plant grows

No cell contents — an empty tube gives little resistance to the flow of water

Rings of lignified cell wall to provide support and prevent collapse

Large empty lumen allows flow of large volumes of water

Thick cell wall — strengthened with lignin to withstand tension and prevent vessel from collapsing

Pits in the cell wall allow water to flow laterally between xylem vessels

Figure 10.2 Xylem vessels showing adaptations

Revision activity

Draw a diagram of a xylem vessel and annotate it with the features that adapt it to transporting water.

Now test yourself

TESTED

1 Explain why the xylem vessel walls are impregnated with lignin and why it is important to have pits in the walls.

Answer on p. 119

Sieve tube elements and companion cells

REVISED

The phloem is adapted to transport assimilates actively by mass flow. There are two cell types involved:

- **sieve tube elements** — long sieve tubes that transport the assimilates
- **companion cells** — support cells that provide all the metabolic functions for the sieve tube elements and are involved in actively loading the sieve tubes

Adaptations of phloem

Phloem cells have several adaptations, as summarised in Table 10.1.

Table 10.1 Adaptations of phloem cells

Cell	Adaptations
Sieve tube elements	Form long tubes
	End walls are retained
	End walls contain many sieve pores, so they are called sieve plates
	Thin layer of cytoplasm
	Very few organelles and no nucleus
Companion cells	Closely associated with sieve tube elements
	Connected to sieve tube elements by many plasmodesmata
	Dense cytoplasm with many mitochondria
	Large nucleus

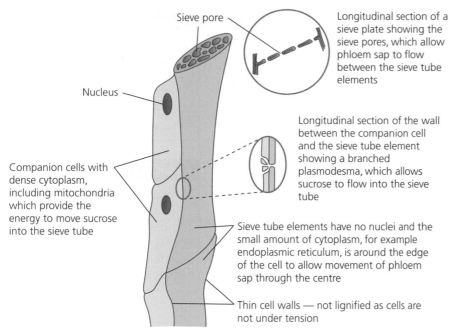

Figure 10.3 Phloem sieve tubes and companion cells showing adaptations

Revision activity

Draw a length of phloem tissue and annotate it with the features that adapt it to transporting assimilates.

Now test yourself

2 Explain why companion cells are essential.

Answer on p. 119

TESTED

The process of transpiration

Transpiration as a consequence of gaseous exchange

REVISED

Transpiration is the loss of water vapour from the upper parts of a plant, mainly the leaves. Although some water evaporates and diffuses through the leaf surface, most water vapour is lost via the stomata. Transpiration involves three stages:

1 Water moves by osmosis from the xylem to the mesophyll cells in the leaf.
2 Water evaporates from the surfaces of the spongy mesophyll cells into the air spaces inside the leaves.
3 Water vapour diffuses out of the leaf via the stomata.

The stomata open during the day to allow gaseous exchange — carbon dioxide enters the leaf and oxygen is released. This is to enable photosynthesis to occur. As the stomata are open, water vapour is lost. Transpiration is therefore a consequence of gaseous exchange.

Factors that affect the rate of transpiration

There must be a water potential gradient between the air spaces in the leaf and the surrounding air to make water vapour leave the leaf. The steeper this gradient, the more rapid the loss of water vapour (transpiration). Factors that increase transpiration rate include:

● higher temperatures — this increases evaporation so there will be a higher water potential inside the leaf
● more wind — this blows water vapour away from the leaf, reducing the water potential in the surrounding air
● lower relative humidity — this increases the water potential gradient between the air inside the leaf and outside
● higher light intensity — this causes the stomata to open wider

Transpiration is the loss of water vapour from the aerial parts of the plant.

Typical mistake

Many candidates state that water is lost from the leaf — it is *water vapour* that is lost.

Revision activity

Draw a cross-section of a leaf and add arrows to show the movement of water molecules. Label each arrow and explain what is happening to make the water to move.

Exam tip

You should explain transpiration using the term *water potential gradient*.

Now test yourself

TESTED

3 Explain why transpiration is quicker on a hot sunny day than on a cool cloudy day.

Answer on p. 119

Answer on p. 119

Measuring the rate of transpiration

Transpiration can be estimated using a bubble **potometer** (Figure 10.4). A potometer actually measures water uptake by the stem, but you can assume that water uptake equals water loss from the leaves in most cases. Care must be taken when setting up the potometer to ensure that there are no leaks and no air in the system, except the bubble used for measuring. Once the shoot has been allowed to acclimatise, the movement of the bubble along the capillary can be measured under different conditions. Transpiration rate is calculated by dividing the distance moved by a set time.

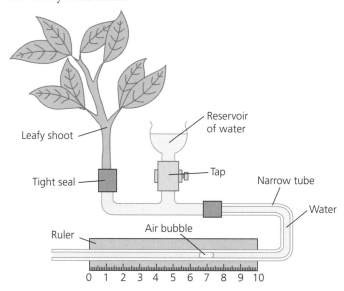

Figure 10.4 A bubble potometer

The transport of water

Transport of water

REVISED

The cell wall of a plant cell is permeable to water. The cell-surface membrane is selectively permeable. As a result, plant cells can contain mineral ions in solution which reduces the **water potential** inside the cell. The more concentrated the mineral ions, the lower the water potential. Water moves by osmosis from a cell with a higher water potential to a cell with a lower water potential because water molecules move down their water potential gradient, as shown in Figure 10.5. The arrows show the direction of water movement between three mesophyll cells in a leaf. Cell P has the highest water potential and the water potential of cell Q is higher than that of cell R.

> **Typical mistake**
>
> Many candidates confuse water potential and water potential gradient. A gradient must be between two points such as between the air inside the leaf and the air outside.

> **Exam tip**
>
> To gain the marks, you must stress that higher temperatures increase transpiration, not simply that temperature increases transpiration.

> **Revision activity**
>
> List the precautions you should take when setting up a bubble potometer.

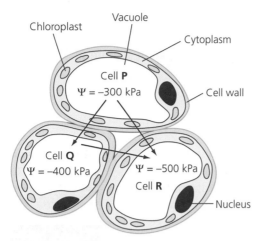

Figure 10.5 Water moves from cell to cell down a water potential gradient

Similarly, water can enter a cell from its environment if the water potential in the cell is lower than the water potential in the environment. This is how root hair cells absorb water from the soil.

Pathways

Once in the plant, water can move across a tissue such as the root cortex by different **pathways** (Figure 10.6):

1 the apoplast pathway carries water between the cells through the cell walls — the water does not enter the cytoplasm or pass through cell-surface membranes
2 the symplast pathway takes water from cell to cell through the cytoplasm of each cell. Water often passes through plasmodesmata linking the cytoplasm of adjacent cells
3 the vacuolar pathway carries water through the cytoplasm and vacuole of each cell

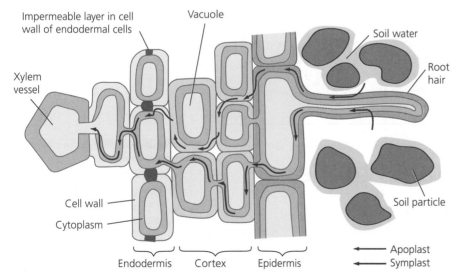

Figure 10.6 Water enters the root hair cells by osmosis. It may pass across the root by the apoplast pathway or the symplast pathway

The transpiration stream

Water movement from the roots up to the leaves in the xylem is known as the **transpiration stream**. There are three mechanisms that move water up the stem: root pressure, adhesion or capillary action and transpirational pull. Root pressure and capillary action combined can only raise water by a few metres. Therefore, transpiration and the pull it creates are essential to move water all the way up a tall stem.

Exam tip

Remember that transpiration is the loss of water vapour from the leaf. The transpiration stream is the flow of water from the roots to the leaves to replace the water lost in transpiration.

The **transpiration stream** is the movement of water from the roots to the leaves.

Now test yourself

4 Explain the difference between transpiration and the transpiration stream.

Answer on p. 119

Root pressure

Root pressure is created by the action of the endodermis in the roots. The endodermis uses metabolic energy to pump mineral ions into the root medulla. This reduces the water potential in the medulla and xylem, making it more negative than in the cortex. Therefore, water moves across the endodermis into the medulla by osmosis.

Water cannot return to the cortex through the apoplast pathway as this is blocked by the Casparian strip. Therefore, pressure builds up in the cortex, which pushes the water up the xylem.

Root pressure is the pressure created by the action of the endodermis.

Typical mistake

Candidates sometimes confuse the movement of water due to hydrostatic pressure differences with the movement between cells caused by a water potential gradient. Water potential gradients cannot move water all the way up the xylem.

Adhesion

Adhesion is the attraction between the water molecules and the walls of the xylem vessel. It results in the water creeping up the xylem in a process called capillary action.

Transpirational pull

Loss of water vapour from the leaves must be replaced by water in the xylem. As water moves out of the xylem, it creates a pull on the column of water in the xylem. As water is cohesive (the molecules attract one another), the column of water is put under tension and pulled up the stem. This is known as the **cohesion-tension theory**.

Adhesion is the attraction between water molecules and the walls of the xylem.

The **cohesion-tension theory** accounts for the movement of water up the xylem.

Typical mistake

Some candidates state that the water moves up the xylem by cohesion tension. This is incorrect — water moves up the xylem as a result of tension created by the loss of water in the leaves, which draws the whole column of water up the xylem due to the cohesion between the water molecules.

Adaptations of plants to the availability of water

Xerophytes

Revision activity

Create a mind map to link together the ideas about how water moves up the xylem.

REVISED

Xerophytes are plants that are adapted to living in dry (or arid) places. The following adaptations help them to reduce loss of water vapour:
- thick waxy cuticle on the leaves
- smaller leaf area
- stomata in pits
- hairy leaves
- rolled leaves

Revision activity

Draw a leaf from a xerophyte and annotate it with all the features that help the plant conserve water. For each feature, explain how it reduces water loss using the terms *water potential* and *water potential gradient*.

Hydrophytes

Hydrophytes are plants that are adapted to living in water, such as water lilies. The following adaptations help them to do this:

1 Leaves and leaf stems have large air spaces (to help them float).
2 The stomata may be on the upper surface of the leaf (to gain carbon dioxide from the air).
3 The stem may be hollow (to allow gases to move to the roots easily).

The mechanism of translocation

Defining translocation

Translocation is the movement of **assimilates** (mostly **sucrose**) around the plant. It occurs in the sieve tubes, but the companion cells are important in actively loading assimilates into the sieve tubes.

Translocation is achieved by mass flow. It is caused by creating a high hydrostatic pressure at the source and a lower hydrostatic pressure at the sink. The fluid in the phloem sieve tube then moves from high to low pressure, i.e. down its pressure gradient.

> **Translocation** is an energy-requiring process for transporting assimilates around a plant.

Sources

A **source** is a part of the plant that has a supply of assimilates that are loaded into the phloem. This could be:

- a leaf that has made sucrose from the products of photosynthesis during the spring and summer
- a root that has stored starch and can convert this to sucrose, which happens particularly in spring
- any other storage organ where the plant has stored starch

> A **source** is a tissue or organ that supplies assimilates to the phloem.

Sinks

A **sink** is a part of the plant that removes sucrose from the phloem and uses or stores it. This could be:

- the buds or stem tips where growth occurs and energy is needed
- the leaves in spring as they grow and unfold
- the roots in summer and autumn when the plant is storing sugars as starch
- any other organ where the plant may store starch

> A **sink** is a tissue or organ that removes assimilates from the phloem and uses them.

Now test yourself

5 Explain how a leaf can be a source or a sink.

Answer on p. 119

Creating high pressure at the source

This process is called **active loading**. Sucrose is moved into the sieve tube by a complex process involving active transport:

1 Hydrogen ions are pumped actively out of the companion cells, which uses ATP as a source of energy.
2 The hydrogen ions diffuse back into the companion cells through special co-transporter proteins carrying sucrose molecules into the companion cells.

> **Exam tip**
>
> The details of the mechanism of translocation are described here in some detail, which is needed to achieve full marks.

3 The sucrose builds up in the companion cells and diffuses into the sieve tube through the many plasmodesmata.
4 The water potential in the sieve tube is reduced.
5 Water flows into the sieve tube by osmosis, increasing the pressure.

Creating lower pressure at the sink

As sucrose is used in respiration or converted to starch in the cells of the sink, the concentration decreases. This creates a concentration gradient between the sieve tubes and the cells in the sink. Sucrose diffuses out of the sieve tubes into the cells and the water potential in the sieve tube increases. Water then moves out of the sieve tube by osmosis.

> **Revision activity**
>
> Draw a flow chart to show how the sieve tube elements are actively loaded.

> **Revision activity**
>
> Write a list of all the key terms used in this chapter, then add the meaning of each key term.

Exam practice

1 Which row in the following table correctly identifies the definitions of the terms given? [1]

Row	Plasmodesmata	Symplast pathway	Casparian strip	Transpiration
A	A connection between cells	Water passes through the cytoplasm	A waterproof layer in the endodermis	Loss of water from the leaves
B	A connection between cells	Water passes along the cell walls	A waterproof layer in the endodermis	Loss of water vapour from the leaves
C	A connection between cells	Water passes through the cytoplasm	A waterproof layer in the endodermis	Loss of water vapour from the leaves
D	A connection between cells	Water passes through the cytoplasm	A waterproof layer in the xylem	Loss of water from the leaves

2 (a) Define the term *transpiration*. [2]
(b) Explain why transpiration is an inevitable result of photosynthesis. [2]
(c) Many xerophytes have a thick waxy cuticle and roll their leaves. Explain how these features reduce transpiration. [5]
3 (a) Define the terms *source* and *sink*. [3]
(b) Name two possible sources. [2]
(c) Explain how a leaf can be a sink and a source at different times. [2]
4 (a) Describe the features of the sieve tubes that help mass flow to occur. [3]
(b) Describe and explain how the companion cells are specialised to their role in loading assimilates into the sieve tubes. [3]

Answers and quick quiz 10 online

ONLINE

Summary

By the end of this chapter you should be able to:
● Explain why large plants need a transport system.
● Describe the distribution of xylem and phloem.
● Describe the structure of xylem and phloem and how they are adapted.
● Define transpiration, describe the factors that affect the rate of transpiration and how to measure the rate using a potometer.
● Explain, using the term *water potential*, how water moves between cells and across plant tissues.
● Describe and explain how water moves up the xylem.
● Describe and explain how assimilates are moved in the phloem.

11 Communicable diseases, disease prevention and the immune system

Pathogens and their transmission

The different types of pathogen

REVISED

Pathogens are microorganisms that cause disease. There is a wide range of them and they are transmitted in a variety of ways (Table 11.1). The means of transmission can be categorised as **direct** or **indirect**.

> A **vector** is an organism that carries the pathogen from one host to another.

Table 11.1 Selected pathogens and their means of transmission

Name of disease	Organism that causes disease	Means of transmission
Athlete's foot (in humans)	Fungus: *Trichophyton rubrum*	Direct contact with **spores** on the skin surface or on other surfaces
Bacterial meningitis	Bacteria: *Neisseria meningitidis* or *Streptococcus pneumonia*	Direct contact with saliva
Black sigatoka (in bananas)	Fungus: *Mycosphaerella fijiensis*	Indirect transmission through spores spread by the wind
Blight (in potatoes and tomatoes)	Fungus-like organism: *Phytophthora infestans*	Direct contact between infected and uninfected seed potatoes; also indirect by spores in the wind
HIV/AIDS	Virus: human immunodeficiency virus	Direct transmission from an infected person by blood to blood contact; infected needles being shared or reused; infected and unsterilised surgical instruments; accidental needle stick; in semen or vaginal fluid during unprotected sexual intercourse; from mother to baby during birth or breast-feeding
Influenza	Virus: from family Orthomyxoviridae — flu viruses	Indirect transmission by droplet infection
Malaria	Protoctistan: *Plasmodium falciparum*, *P. vivax*, *P. ovale*, *P. malariae*	Indirect transmission via a vector — the vector is the female *Anopheles* mosquito
Ring rot (in potatoes and tomatoes)	Bacteria: *Clavibacter michiganensis* subsp. *sepedonicus*	Direct contact between infected and uninfected potatoes
Ringworm (in cattle)	Fungus: *Trichophyton verrucosum*	Direct contact with spores on the skin surface or on other surfaces
Tobacco mosaic virus	Virus: tobacco mosaic virus	Direct contact between infected and uninfected leaves; also indirect by aphids acting as **vectors**
Tuberculosis (TB)	Bacteria: *Mycobacterium tuberculosis* and *M. bovis*	Indirect transmission by bacteria carried in tiny water droplets in the air

TESTED

Now test yourself

1 Explain how a vector can transmit a pathogen without being infected by the pathogen itself.

Answer on p. 119

Plant defences against pathogens

REVISED

Organisms are surrounded by pathogens and have evolved defences against them. Plant defences can be divided into physical and chemical.

Physical defences

Cellulose cell walls, lignin thickening of cell walls, waxy cuticles and bark all help to prevent entry of a pathogen. Inside the plant, old vascular tissue is blocked to prevent a pathogen spreading easily. **Callose** is deposited around the sieve plates in older sieve tubes and blocks flow. Tylose (a balloon-like swelling) also blocks old xylem vessels.

When a pathogen is detected, the following barriers can be enhanced to prevent it spreading through the plant:

1 The stomata close to prevent entry to the leaves.
2 Cell walls are thickened with additional cellulose.
3 Callose is deposited between the plant cell wall and cell membrane near the invading pathogen. This strengthens the cell wall and blocks the plasmodesmata.
4 Necrosis (deliberate cell 'suicide') — cells surrounding the infection are killed to reduce access to water and nutrients. Necrosis is caused by intracellular enzymes that are activated by injury. These enzymes kill damaged cells and produce brown spots on leaves.
5 Canker — necrosis of woody tissue in the main stem or branch.

> **Callose** is a large polysaccharide that blocks old sieve tubes.

Now test yourself

2 Explain why blocking the phloem with callose and the xylem with tylose reduces the spread of a pathogen.

Answer on p. 119

TESTED

Chemical defences

Chemicals such as terpenes and tannins are present to prevent entry of a pathogen. However, other chemicals can be released when a pathogen is detected. Chemical defences include:

1 terpenoids — essential oils with antibacterial and antifungal properties, e.g. menthols produced by mint plants
2 phenols (e.g. tannin) — found in bark, which has antibiotic and antifungal properties. Tannins inhibit insect attack by interfering with their digestion
3 alkaloids (e.g. caffeine, nicotine, cocaine and morphine) — give a bitter taste to inhibit herbivores feeding. They also interfere with metabolism by inhibiting or activating enzyme action
4 defensins — small cysteine-rich proteins that inhibit ion transport channels in the plasma membrane of pathogen cells
5 hydrolytic enzymes — found in the spaces between cells. These include chitinases, which break down the chitin in fungal cell walls; glucanases, which hydrolyse the glycosidic bonds in glucans of bacterial walls; and lysosymes, which degrade bacterial cell walls.

Animal defences against pathogens

Animal defences against pathogens take two forms:
1 **primary defences**, which prevent entry of pathogens into the body
2 **secondary defences**, which combat pathogens that have already entered the body

Primary defences

The skin is the main primary defence against pathogens and parasites. It provides a barrier to the entry of microorganisms.

Blood clotting and skin repair

Damage to the **skin** opens the body to infection. **Blood clots** reduce the loss of blood and make a temporary seal, preventing access by pathogens. Blood-clotting involves calcium ions and at least 12 other clotting factors that are released by platelets or from the damaged tissue. These factors activate an enzyme cascade, which produces insoluble fibres (Figure 11.1). The clot dries out to form a scab. Over time, the scab shrinks, drawing the sides of the cut together. Fibrous collagen is deposited under the scab and stem cells in the epidermis divide to form new cells. These cells differentiate to form new skin.

> **Primary defences** prevent the entry of the pathogen into the body.
>
> **Secondary defences** help to remove a pathogen after it has entered the body.

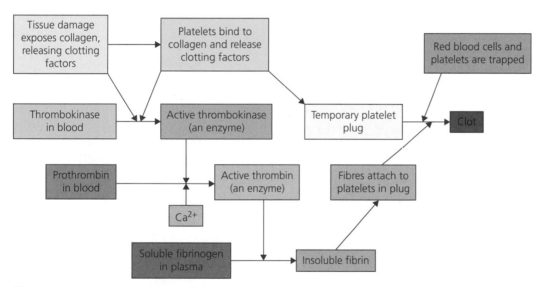

Figure 11.1 An enzyme cascade causes blood to clot

Inflammation

Inflammation is the swelling and redness seen in infected tissues. Infected tissue often feels hot and tender. This is caused by a **cell-signalling** substance called histamine. Histamine is released from mast cells and it has a range of effects that act to help combat the infection. It causes vasodilation and makes the capillary walls more permeable to white blood cells and some proteins. More plasma enters the tissue fluid, causing oedema (swelling). The excess tissue fluid is drained into the lymphatic system, moving the infecting pathogens towards the lymph nodes where lymphocytes can initiate the specific **immune response**.

Mucous membranes

Any areas where the skin is incomplete are protected by **mucous membranes**. This includes the airways, lungs, digestive system, ears and

Now test yourself

3 Explain why the blood-clotting process needs to be so complex.

Answer on p. 119 TESTED

> The **immune response** is the body's response to invasion by pathogens.

genital areas. The epithelial layer contains mucus-secreting goblet cells. The mucus traps any pathogens, immobilising them. Some areas, such as the airways, also have ciliated cells. Cilia are tiny, hair-like organelles that can move in a coordinated fashion to waft the layer of mucus along.

Mucous membranes are sensitive to irritation. They respond to the presence of microorganisms or the toxins they release. This causes expulsive reflexes (e.g. coughing, sneezing and vomiting). In a cough or sneeze, the sudden expulsion of air carries with it the microorganisms causing the irritation.

Secondary defences

The non-specific immune response

Phagocytosis

This involves phagocytic white blood cells (**neutrophils**) that engulf and destroy any non-self cells. Neutrophils are cells that contain a lobed nucleus and dense cytoplasm that contains many **lysosomes** and mitochondria. They also have a well-developed cytoskeleton.

Non-self cells are recognised because they have proteins on their cell-surface membranes called **antigens**. They may also be identified by the presence of opsonins, which are non-specific **antibodies** that bind to pathogens and act as binding sites for the phagocyte.

Phagocytosis follows a particular sequence (Figure 11.2):
1 The bacteria are engulfed by the neutrophil.
2 They are surrounded by a vacuole called a **phagosome**.
3 Lysosomes fuse to the phagosome.
4 Lytic enzymes are released into the phagosome.
5 The bacteria are hydrolysed (digested) and the nutrients can be reabsorbed into the cell.

Antigen presentation

Some phagocytes (macrophages) can engulf the pathogen and process it so that the antigen is kept whole. This antigen can then be placed in a special protein complex on the cell-surface membrane. This is known as **antigen presentation**. These **antigen-presenting cells** can then be used to initiate the **specific immune response**.

The specific immune response

The specific response relies on the action of the **B lymphocytes** and **T lymphocytes**, which are cells with a large nucleus. They have specific receptors on their cell-surface membranes. These receptors show a wide diversity of shapes, each of which is complementary to the shape of an antigen on a specific pathogen.

The specific response is more complex than the non-specific response and involves a series of stages (Figure 11.3).
1 *Antigen presentation.*
2 *Clonal selection.* There may be only a few lymphocytes that carry the correct receptors to bind with the specific antigen. Normally, there is a B lymphocyte, a **T helper** lymphocyte and a **T killer** lymphocyte. These cells must be found and activated. They are selected and activated by coming into contact with the specific foreign antigen. This ends the cell-signalling role of the antigen presenting cell. The action of macrophages makes this selection more likely.

Antigens are molecules on the surface of cells that the immune system can use to recognise pathogens.

Antibodies are proteins that are secreted in response to stimulation by the appropriate antigen. They have specific binding sites and are capable of acting against the pathogen.

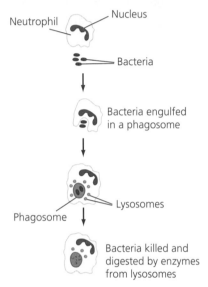

Figure 11.2 The stages involved in phagocytosis

Antigen presentation involves placing an antigen on the cell-surface membrane of a phagocytic cell.

The **specific immune response** means that the response deals only with one pathogen that possesses one particular antigen.

Exam practice answers and quick quizzes at **www.hoddereducation.co.uk/myrevisionnotes**

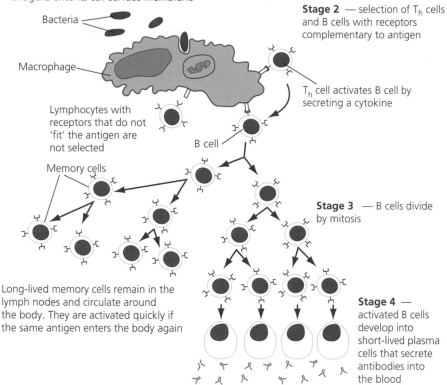

Stage 1 — macrophage engulfs bacterium and digests it, putting bacterial antigens onto its cell-surface membrane

Bacteria

Macrophage

Lymphocytes with receptors that do not 'fit' the antigen are not selected

Memory cells

B cell

Stage 2 — selection of T_h cells and B cells with receptors complementary to antigen

T_h cell activates B cell by secreting a cytokine

Stage 3 — B cells divide by mitosis

Long-lived memory cells remain in the lymph nodes and circulate around the body. They are activated quickly if the same antigen enters the body again

Stage 4 — activated B cells develop into short-lived plasma cells that secrete antibodies into the blood

Figure 11.3 The stages of the specific immune response

3 *Clonal expansion and differentiation.* Once activated, the specific lymphocytes increase in numbers by mitosis. There will normally be a clone of B lymphocytes, a clone of T helper lymphocytes and a clone of T killer lymphocytes. Cells from the clone of B lymphocytes will differentiate into plasma cells and **memory cells**. The plasma cells are short-lived. They produce and release the antibodies which are proteins that have specific binding sites and are capable of acting against the pathogen.

4 *Differentiation.* Cells from the T helper lymphocyte clone will differentiate into T helper cells and memory cells. The T helper cells secrete hormone-like substances called **cytokines** and **interleukins** that help to stimulate the B lymphocytes and the macrophages. Memory cells are long-lived and remain in the blood for some time, providing immunological memory or long-term immunity. If the same pathogen invades again, it will be recognised and attacked more quickly. The T killer clones manufacture harmful substances and attack and kill host cells that are already infected. **T regulator** cells have a role in closing down the immune response once the pathogen has been removed.

> **Typical mistake**
>
> Many candidates still seem to confuse antigens and antibodies.

> **Exam tip**
>
> To gain full marks, you must be specific — antibodies are made by the plasma cells.

> **Revision activity**
>
> Try to visualise this whole process by drawing a simple flow diagram.

Now test yourself

TESTED

4 Explain why pathogens have antigens that allow them to be recognised as foreign on their surface.

Answer on p. 119

Primary and secondary immune responses

Primary responses

The first time a pathogen invades the body, it produces a **primary response.** This takes a few days as the specific B and T lymphocytes must be selected, the cells must divide and then differentiate and the antibodies must be manufactured. As a result, the peak of activity and the maximum concentration of antibodies are not achieved until several days after infection.

> The **primary response** is the immune system's response to a first infection.

Secondary responses

On any subsequent invasion by the same pathogen with identical antigens on its surface, the immune response is a **secondary response**. The immune system can respond much more quickly and with higher intensity — it is much more effective. This is because the blood carries many memory cells that are specific to this pathogen. They divide and differentiate into plasma cells, which then manufacture antibodies. The response is quick enough to prevent the pathogen taking hold and causing symptoms of illness.

> The **secondary response** is the immune system's response to a second or subsequent infection by the same pathogen.

The structure and general functions of antibodies

The general structure

Antibodies are large globular proteins. Most antibodies are similar in structure and possess the same basic components. They consist of four polypeptides held together by disulfide bridges (Figure 11.4).

One end of an antibody is known as the constant region. This has a binding site that can be recognised by phagocytes. It may help binding of phagocytes. The opposite end is known as the variable region. This has binding sites that are specific to a particular antigen. They have a shape that is complementary to the shape of the antigen and can bind to that antigen. The antibody molecule also has a hinge region that allows some flexibility to enhance binding to more than one pathogen.

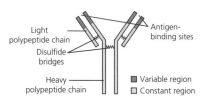

Figure 11.4 The structure of an antibody molecule

The action of antibodies

Antibodies can act on pathogens in a number of ways. These include:
1 **opsonins** — antibodies that bind to a pathogen and help phagocytes to bind
2 **agglutins** — antibodies that bind to two or more pathogens so that they are held together. This prevents them entering cells and reproducing. Some antibodies have many binding sites. These can be used to bind to a number of pathogens.
3 **anti-toxins** — antibodies that bind to the toxin, making it harmless

Types of immunity

Active and passive immunity

Active immunity is acquired through activation of the immune system. It involves the selection of specific lymphocytes and the production of antibodies and memory cells. The memory cells remain in the blood for a long time, providing long lasting immunity.

Passive immunity is acquired from another source. Antibodies may be injected straight into the blood or acquired from a mother's milk. These antibodies do not last long, but they do give immunity from a specific pathogen for a period of time. No memory cells are made, so the immunity is not permanent.

> **Active immunity** is immunity acquired by activation of the immune system.
>
> **Passive immunity** is when someone is given antibodies produced by someone else.

Natural and artificial immunity

Natural immunity is acquired in the course of everyday activity. Natural active immunity may be the result of catching a flu virus from someone who sneezes. Natural passive immunity may be acquired through breast milk or the mother's placenta.

Artificial immunity is acquired by human intervention. Artificial active immunity may be the result of a vaccination. Artificial passive immunity results from an injection of antibodies.

Autoimmune diseases

Defining autoimmune diseases

An **autoimmune disease** is one in which the immune system attacks the body's own healthy cells and tissues. The B and T lymphocytes usually respond to antigens on harmful organisms such as bacteria or viruses. In an autoimmune disease, the lymphocytes do not distinguish between these 'foreign' antigens and your own. As a result, they release antibodies that attack your own tissues.

This attack may be due to the exposure of antigens that are not normally exposed, such as certain molecules usually found only in the nucleus. The exact cause of autoimmune disease is not known, but it may be due to changes triggered by drugs or by an infection with certain bacteria or viruses. It also seems that certain people are more genetically prone to autoimmune disease.

An autoimmune disease may cause:
1 damage to body tissues and organs
2 abnormal growth of an organ
3 changes in organ function

Examples include **arthritis** (painful inflammation of joints) and **lupus**, which can affect any part of the body causing swelling and pain.

The principles of vaccination

Routine vaccinations

Vaccination is the deliberate introduction of antigenic material in order to stimulate the production of antibodies. Antigenic material can be:
- whole live microorganisms (this is more effective than using dead ones as the live organisms can reproduce and mimic an infection better)

- dead microorganisms
- attenuated (weakened) organisms
- a surface preparation of antigens
- a toxoid (a harmless form of a toxin)

Herd and ring vaccination

In order to be effective, vaccinations need to be used appropriately. Herd vaccination involves the systematic vaccination of all or most members of a population. This prevents the pathogen being transmitted from person to person. Ring vaccination is a response to an outbreak. All the people in the area surrounding the outbreak are vaccinated in order to prevent transmission and to isolate the outbreak in one area.

Possible sources of medicines

New medicines

Scientists are always looking for new medicines to help combat disease. **Microorganisms** and **plants** produce a wide range of molecules that may be of benefit in fighting disease. Among the huge diversity of plants and microorganisms in the natural world there may be organisms that produce chemicals that are beneficial against a wide range of pathogens or in fighting diseases such as cancer. We just need to find them.

Every species of organism that is allowed to become extinct could potentially hold the cure for a major disease. It is therefore important that we try to maintain biodiversity and conserve as many species as possible, just in case the molecules they are capable of producing may prove to be useful.

Personalised medicine

Screening the genomes of many plants and microorganisms enables scientists to identify compounds that may have a medicinal value. As sequencing technology and molecular modelling techniques improve, it may be possible to sequence the DNA of an individual to assess a specific genetic disorder and develop a treatment personalised to the individual, such as an individualised combination of drugs. Such drugs could be those identified from plants or microorganisms, or these compounds could be modified to become more effective. In the future, it may even become possible to design a drug that is specific to the needs of one individual.

Synthetic biology

Scientists can modify natural molecules for use as medicines. Synthetic biology is the development of new molecules or systems. This could be making new enzymes to create a new product or using natural enzymes to produce a new effect. For example, bacteria are being genetically modified for use as biosensors. Scientists have created new pathways that enable bioluminescent bacteria to release light in response to specific pollutants.

Benefits and risks of antibiotics

The wide use of antibiotics

The first antibiotic, **penicillin**, was discovered by Alexander Fleming in 1928. Since then, many different antibiotics have been discovered and used widely to treat bacterial infections.

The use of antibiotics became widespread during the Second World War to prevent the infection of wounds, which has saved many lives. However, their misuse has enabled microorganisms to develop resistance. **Clostridium difficile** (*C. diff*) and methicillin-resistant *Staphylococcus aureus* (**MRSA**) are well known for their multiple **resistance** to antibiotics.

Exam practice

1 Which row in the following table shows the correct definitions of the terms given? [1]

Row	Antigen	Antibody	Pathogen	Vaccination
A	A protein made by the immune system	A molecule on the surface of a pathogen	A microorganism that causes disease	The introduction of antigenic material
B	A molecule on the surface of a pathogen	A protein made by the immune system	A microorganism that causes disease	The introduction of antigenic material
C	A molecule on the surface of a pathogen	A protein made by the immune system	An organism that lives on another orgainism	The introduction of antigenic material
D	A molecule on the surface of a pathogen	A protein made by the immune system	A microorganism that causes disease	An injection

2 (a) Describe three ways in which HIV is transmitted. [3]
   (b) The following table shows the percentage of people with new HIV infections in four regions.

	Percentage of people with new HIV infections			
Year	Western Europe	Eastern Europe	Far East	Sub-Saharan Africa
1980	0.0	0.0	0.0	0.0
1990	0.1	0.2	0.1	2.0
2000	0.4	0.5	0.3	8.5
2010	0.3	0.9	0.3	15.2

   (i) Describe the trends shown by the data in the table. [4]
   (ii) Suggest why the trend in Western Europe is different from that in other areas. [3]
   (c) Discuss the advantages and disadvantages of treating HIV/AIDS patients with antibiotics. [4]
3 (a) Explain the difference between antigens and antibodies. [3]
   (b) Describe the structure of an antibody and explain how it is adapted to its function. [4]
   (c) Suggest and explain why people in underdeveloped parts of the world such as sub-Saharan Africa may be unable to make sufficient antibodies. [4]

Answers and quick quiz 11 online

ONLINE

Summary

By the end of this chapter you should be able to:
- Describe the types of pathogen that cause disease and how they are transmitted.
- Describe plant defences against pathogens.
- Describe the primary defences against pathogens in animals.
- Describe the action of phagocytes.
- Describe the structure and action of B and T lymphocytes.
- Compare and contrast primary and secondary immune responses.
- Describe the structure and action of antibodies.
- Compare and contrast active, passive, natural and artificial immunity.
- Explain how vaccination can control disease.
- Outline the possible sources of new medicines.
- Outline the benefits and risks of antibiotics.

12 Biodiversity

The levels of biodiversity

> **Exam tip**
>
> This topic covers key scientific terms with precise meanings. You should learn them and use them only in the correct context.

Considering biodiversity at different levels

REVISED

Biodiversity is the variety of life. It includes all the different plant, animal, fungus and microorganism species in the world, the genes they contain and the ecosystems of which they form a part. Biodiversity can be considered at three levels: the range of **habitats** within an ecosystem, the range of **species** within a habitat and the **genetic** variation within a species (different breeds within a species).

> **Revision activity**
>
> Draw a mind map with biodiversity in the centre. Include the three levels of biodiversity and explain what they mean.

> **Biodiversity** is the variety of life on Earth.
>
> A **habitat** is the home or environment of an animal, plant or organism.
>
> A **species** is a group of organisms with similar adaptations that live and breed together to produce fertile offspring.

Sampling

REVISED

Most habitats are large in area and have large numbers of plants and animals. It is therefore impossible to count the number of individuals in each species. **Sampling** involves studying small parts of the habitat in detail and then multiplying up to calculate the size of a population. It is assumed that the sample plots are representative of the entire habitat.

When sampling, it is important to consider the:
- size of the samples — this depends on the size of the habitat
- number of sample areas used — the more sample areas used the better, as the results will be more reliable
- sampling technique used — this must be identical in every sample

The sampling should not disturb the habitat more than is essential.

Techniques

There are two main techniques for sampling: random and non-random.

Random sampling avoids bias. It can be achieved by using a computer to generate random numbers, which are then used as coordinates to locate sample areas on an imaginary grid placed over the habitat.

Non-random sampling includes:
1 **Opportunistic sampling**, which involves using prior knowledge to select sample sites or changing the sampling strategy once onsite.
2 **Stratified sampling**, which involves carrying out samples in each recognisable sub-habitat.
3 **Systematic sampling**, which involves carrying out sampling at fixed intervals in each direction.

> **Now test yourself**
>
> 1 Explain why it is important to avoid bias in sampling.
>
> **Answer on p. 119**
>
> TESTED

Methods

The method of sampling used depends on the type of vegetation in the habitat and what type of organisms are being examined. It is important to measure both the number of species (species richness) and the number of individuals in each species (species evenness).

Plants

Large plants such as trees can be counted individually. Smaller plants can be sampled using quadrats. These are square frames of a suitable size that are placed over a random site and examined closely to identify all the plants inside the quadrat. A quantitative sample can be achieved by measuring the percentage cover of each species within the quadrat. This can be done using the following methods:

- point sampling — place a point frame in the quadrat and count the number of examples of each species that touch each point
- grid sampling — divide the quadrat using string into a known number of smaller squares (often 100) and then estimate how many squares are occupied by each species

Animals

Large animals can be sampled by careful observation and counting. Smaller animals will need to be caught or trapped.

Small mammals can be trapped using a humane trap such as a Longworth trap. It is possible to estimate the population size by the mark and recapture technique. This involves two separate trapping sessions. The animals caught the first time are marked in a way that causes them no harm. If the number of animals trapped in the first session is T1, the number caught in the second session is T2 and the number caught in the second session that are already marked is T3, the total population can be given by the formula:

$$\text{Number in population} = T1 \times \frac{T2}{T3}$$

Ground-living invertebrates can be collected using a pitfall trap, whereas invertebrates in leaf litter can be collected using a Tullgren funnel. Invertebrates in trees can be collected by using a stick to knock a branch and collecting in a sheet the organisms that fall to the ground. Invertebrates in grass and shrubs can be collected by sweep netting, and pond life can be sampled by netting.

Species richness and species evenness

Measuring species richness and species diversity

REVISED

A count of the number of species in a habitat is called the **species richness**. However, a habitat that is dominated by just one species with only one or two individuals of each of the other species would not be considered to be biologically diverse. A habitat in which all species are equally represented is more diverse. This is known as **species evenness**. Therefore, a measurement of the biodiversity of a habitat should account

> **Species richness** is the number of different species in a habitat.
>
> **Species evenness** is how evenly each species is represented throughout a habitat.

for both the number of species and the number of individuals in each species living in that habitat. A diverse habitat would contain a large number of species, all of them represented by a sizeable population rather than by just one or two individuals.

Now test yourself   TESTED ☐

2 Explain the significance of species richness and species evenness.

Answer on p. 119

Simpson's Index of Diversity

Understanding and using the formula   REVISED ☐

Simpson's Index of Diversity measures the biodiversity of a habitat. There are several versions of the formula, so make sure you are consistent. The most commonly used version is:

$$D = 1 - (\Sigma(n/N)^2)$$

where n = the total number of individuals in a particular species and N = the total number of individuals in all species.

If it is not possible to count all the individual plants in an area, percentage cover can be used. The resultant value always ranges between 0 and 1.

Worked example

The data collected from a field may look like this those in Table 12.1. To measure Simpson's Index of Diversity, further calculations are carried out (Table 12.2).

Table 12.1 **Data collected from field sampling**

Species	Percentage cover
Yorkshire fog	16
Meadow grass	74
Bent grass	2
Thistle	3
Buttercup	4
Dock	1

Table 12.2 **Calculations for Simpson's Index of Diversity**

Species	Percentage cover	n/N	$(n/N)^2$
Yorkshire fog	16	0.16	0.0256
Meadow grass	74	0.74	0.5476
Bent grass	2	0.02	0.0004
Thistle	3	0.03	0.0009
Buttercup	4	0.04	0.0016
Dock	1	0.01	0.0001
Total	100	1.00	$\Sigma = 0.5762$

Typical mistake

Some candidates don't find calculations easy — here they often forget to square each number or to subtract from 1 as the final step in the calculation.

Therefore, using Simpson's Index of Diversity:

$D = 1 - (\Sigma(n/N)^2)$

$D = 1 - (0.0256 + 0.5476 + 0.0004 + 0.0009 + 0.0016 + 0.0001)$

$= 1 - 0.5762$

$= 0.42$

Exam tip

You may be asked to calculate Simpson's Index of Diversity from data provided. Calculations involving the index are not as difficult as many students imagine. The formula and a partly completed table are often provided.

Interpretations of high and low values

A **high value** (close to 1) indicates:
- a habitat with high diversity
- there are a good number of species (high species richness)
- the species are relatively evenly represented (high species evenness)
- the habitat should be stable and may survive some disruption
- the habitat is probably one that is worth conserving

A **low value** (close to 0) indicates:
- a habitat with low diversity
- the habitat is dominated by one or a few species
- the habitat may be unstable and damaged by disruption
- the habitat may be manmade

In the worked example above, a diversity index of 0.42 is not particularly high. This habitat is dominated by one species (meadow grass). If this species were harmed by human action, the habitat may be unstable.

Now test yourself

3 Explain why an ecologist might want to monitor the diversity of a habitat.
4 Explain why a conservationist may want to conserve a more diverse habitat.

Answers on p. 120

TESTED ☐

Genetic biodiversity

Assessing genetic diversity

REVISED ☐

The **genetic diversity** of a population may be important to conservationists wishing to maintain the health of a captive population in a zoo or rare-breed centre. It may also be important in pedigree charts. Genetic diversity is increased when there is more than one **gene variant** (**allele**) for a particular **locus**. If there are several alleles, it is called a **polymorphic gene locus**. A measure of genetic diversity is given by the formula:

$$\text{Proportion of polymorphic gene loci} = \frac{\text{number of polymorphic gene loci}}{\text{total number of loci}}$$

The factors affecting biodiversity

REVISED ☐

Many species have become rare or endangered, often because of human activity. The main factors that affect biodiversity include:

1 the rapid **human population growth**, which means that more space and resources are taken up to supply living space and food
2 the increasing use of **agriculture** (**monoculture**) as an efficient way to produce food and other products. This decreases biodiversity and reduces the size of natural habitats, which may become unstable as a result
3 the increasing release of waste products that pollute the atmosphere, causing **climate change**

Maintaining biodiversity

The reasons for maintaining biodiversity

It is important to conserve endangered species for a variety of reasons, many of which can be applied generally to most endangered species.

Ecological reasons

Many species are in decline because of habitat destruction — finely balanced ecosystems may be disrupted by a small change. This is particularly true where **keystone species** are involved. For example, beavers make dams that cause flooding and introduce new conditions in which many aquatic species live. All those species would be lost if beavers were not present.

The whole balance of life on Earth is maintained by the activity of species within ecosystems and we do not know what knock-on effects the loss of one species may have. The **ecological reasons** for maintaining the correct functioning of ecosystems include:
- fixing of energy from sunlight
- regulation of the oxygen and carbon dioxide levels in the atmosphere
- fresh water purification and retention
- soil formation
- maintenance of soil fertility
- mineral recycling
- waste detoxification and recycling

> A **keystone species** is one that has a disproportionate effect on the ecosystem — the loss of one species may affect many others.

> **Exam tip**
>
> The reasons for conservation are generic, but questions in the examination are likely to ask about a specific case. Be ready to apply these generic ideas to the case under investigation.

Economic reasons

Economic reasons for conserving species include:
- Growth of food and timber relies on the correct functioning of ecosystems. Even the soil relies on the ecosystem functioning correctly. If an ecosystem is disrupted, the effects may be far-reaching. An unbalanced ecosystem could cause **soil depletion** so that it loses fertility. This is particularly obvious where monoculture is used constantly.
- Pollination of many crops relies on insects, particularly bees.
- Natural predators to pests reduce the need for pesticides.
- As yet unknown species may contain molecules that are effective medicines.

> **Typical mistake**
>
> Many students write long, heartfelt responses about the right of all organisms to live and how humankind should not 'play God'. This sort of response may gain some credit, but no more than 1 or 2 marks.

Aesthetic reasons

Aesthetic reasons for maintaining biodiversity are important:
- Everyone enjoys nature or its benefits in one way or another. A healthy, well-balanced ecosystem with its variety of life forms, colours and activity is complex and beautiful.
- Being surrounded by natural systems relieves stress and helps recovery from injury.
- Maintaining the landscape.

In situ and *ex situ* conservation

In situ conservation

In situ **conservation** involves conserving a species in its natural habitat by creating **marine conservation zones** and **wildlife reserves**. Table 12.3 outlines the advantages and disadvantages of this form of conservation.

Table 12.3 The advantages and disadvantages of *in situ* conservation

Advantages	Disadvantages
The organisms are in their normal environment	It can be difficult to monitor the organisms and ensure that they are healthy
The habitat is conserved, along with all the other species living in it	The environmental factors that caused the decline in numbers may still be present
The organisms will behave normally	Poaching or hunting may continue
It generates work for local people looking after the reserve	There may be food shortages
Ecological tourism can generate income	Disease will be difficult to treat
	Predators can be difficult to control

Ex situ conservation

Ex situ **conservation** involves conserving a species using controlled habitats away from its normal environment. **Seed banks**, **botanic gardens** and **zoos** all keep groups of individuals of endangered species. Table 12.4 outlines the advantages and disadvantages of this form of conservation.

Table 12.4 The advantages and disadvantages of *ex situ* conservation

Advantages	Disadvantages
Research is easy	The organisms are living in an unnatural habitat
The organisms can be monitored to ensure they remain healthy	The organisms may not behave as normal
The organisms can be kept in separate populations to ensure a disease does not affect the whole species	The organisms may not breed
Breeding can be controlled to prevent in-breeding and subsequent loss of genetic diversity	There is little point in conserving individuals if their natural habitat is lost and there is nowhere for them to return to
Genetic diversity can be increased by sharing specimens with other conservation sites	
Once populations have been increased, individuals can be reintroduced to the wild	
Seed banks can store the seeds of millions of rare plant species or plant species that are extinct in the wild, using the same area necessary for only a few hundred adult plants	

International and local conservation agreements

REVISED

For conservation activities to be effective, it is essential that all parties agree on what must be done. A wide range of agreements are aimed at ensuring that conservation efforts are successful.

Typical mistake

Many candidates can answer questions on this topic in generic terms but tend to get thrown by specific examples.

Exam tip

The principles of conservation are straightforward. Therefore, questions about conservation are likely to be applied to a specific case. Learn to apply your knowledge to new contexts.

Revision activity

Look at a local wildlife park's website to find out what activities are carried out for international conservation.

Now test yourself

5 Explain why it is important to keep more than one population of an endangered species.

Answer on p. 120

TESTED

The Convention on International Trade in Endangered Species (CITES)

The **Convention on International Trade in Endangered Species (CITES)** is an international agreement between governments, to which countries adhere voluntarily. Its aim is to ensure that international trade in specimens of wild animals and plants does not threaten their survival.

The Rio Convention on Biological Diversity (CBD)

The **Rio Convention on Biological Diversity (CBD)** recognises that people need to secure resources of food, water and medicines. However, it promotes development that is sustainable and partner countries agree to adopt *ex situ* conservation measures with shared resources.

The Countryside Stewardship Scheme (CSS)

The **Countryside Stewardship Scheme (CSS)** aims to enhance the natural beauty and diversity of the UK's countryside and improve public access. This includes looking after wildlife habitats, retaining the traditional character of the land and protecting historic features and natural resources.

It applies to land that is not considered to be an environmentally sensitive area. Payments are made to landowners to manage the land in a suitable manner and for capital works such as hedge laying, planting and repairing dry-stone walls.

> **Revision activity**
>
> Write a list of all the key terms used in this chapter, then add the meaning of each key term.

Now test yourself

TESTED

6 Explain why international cooperation is essential for successful conservation.

Answer on p. 120

Exam practice

1 (a) State what is meant by the terms *biodiversity*, *species richness* and *species evenness*. [3]

  (b) Biodiversity can be measured at three levels: habitat, species and genetic. Explain what is meant by each level and explain the significance of high diversity in each case. [6]

  (c) A student collected the following data from two fields.

Species	Percentage cover	
	Field A	Field B
Rye grass	76	14
Bent grass	0	84
Fescue	45	0
Dandelion	9	0
Buttercup	24	2
Daisy	18	0
Total	172	100

    (i) Calculate the value of Simpson's Index of Diversity for field B. [2]

    (ii) The student calculated Simpson's Index of Diversity for field A. The value was 0.704. Suggest what the relative values for fields A and B mean about their diversity and their value as habitats. [4]

2 The mountain gorilla (*Gorilla beringei beringei*) is an endangered species. There are thought to be about 790 individuals left in the wild. These live in two populations, one in the Virunga mountains of central Africa and the other in southwest Uganda.

  (a) Explain why such a low population puts the species at risk. [3]

  (b) (i) Suggest how local people could help to conserve these gorillas. [3]

     (ii) Suggest how *ex situ* conservation techniques could be used to help the gorillas. [3]

3 Explain why seed banks are considered to be more important than botanical gardens. [2]

4 (a) What do the initials CITES mean? [1]

  (b) (i) Describe the aims of CITES. [2]

     (ii) Suggest two ways in which these aims can be achieved. [2]

  (c) Describe the role of the Countryside Stewardship Scheme. [3]

Answers and quick quiz 12 online

ONLINE

Summary

By the end of this chapter you should be able to:
- Explain how biodiversity can be considered at the level of habitat, species and genetic.
- Explain the importance of sampling when measuring the biodiversity of a habitat.
- Describe how samples can be taken.
- Describe how to measure biodiversity in a habitat.
- Use Simpson's Index of Diversity.
- Outline the significance of high and low values of the Simpson's Index of Diversity.
- Describe how genetic diversity can be estimated.
- Outline the reasons for the loss of biodiversity.
- Describe the conservation of endangered plant and animal species, both *in situ* and *ex situ*, with reference to the advantages and disadvantages of these two approaches.
- Discuss the importance of agreements to the successful conservation of species with reference to CITES, the Rio Convention on Biodiversity and the Countryside Stewardship Scheme.

13 Classification and evolution

Biological classification

Taxonomic hierarchy

REVISED

Biologists use **biological classification of species** to place living things into groups to make them easier to study. These groups are called taxonomic groups or taxa (singular: taxum). The **taxonomic hierarchy** is shown in Figure 13.1. Therefore, **domain** is the largest group (or taxum) and **species** the smallest. Similar species are placed in a **genus**. Similar genera are placed in a **family** etc.

Biological classification of species is placing living things into groups.

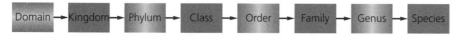

Figure 13.1 The hierarchy of biological classification

Now test yourself

1 Explain why the order of taxonomic groups is known as a hierarchy.

Answer on p. 120

TESTED

The binomial system

The system and its advantage

REVISED

The **binomial system** is the way in which we use two Latin words to name each species. The first name is the name of the genus to which the species belongs and the second name is the specific or species name. For example, in the term *Homo sapiens*, *Homo* is the genus of man and *sapiens* is our species name. Both of the Latin names should always be written in italics or underlined. The genus name should be written with an upper case first letter, whereas the species should be in lower case.

The main **advantage** of this system is to avoid the confusion that can arise if local names are used. The binomial system is recognised worldwide.

The features used to classify organisms

The five kingdoms

REVISED

Carl Linnaeus set up the first true system of classification system, which is still used today. He used the organisms' **observable features** and grouped them according to the number of **similarities** present.

More recently, taxonomists study organisms in ever-more detail, describing all of their observable features. They then decide if any differences are simply a variation within the species or significant enough to classify the organism as a different species.

Typical mistake

Candidates often use the binomial system incorrectly, forgetting to capitalise the genus name or not using italics for the genus and species names.

In the **five kingdoms** system of classification, the kingdoms are separated by characteristic features that are summarised in Table 13.1.

Table 13.1 The characteristic features of the five kingdoms

| Feature | Kingdom | | | | |
	Prokaryotae	Protoctista	Fungi	Plantae	Animalia
Cellular	Unicellular	Unicellular, some multicellular (algae)	Acellular (body composed of mycelium). Yeasts are unicellular	Multicellular	Multicellular
Nucleus	No	Yes	Yes (cytoplasm is multinucleate)	Yes	Yes
Membrane-bound organelles	No	Yes	Yes	Yes	Yes
Cell wall	Yes (made of peptidoglycan)	Present in many species	Yes (made of chitin)	Yes (made of cellulose)	No
Nutrition	Autotrophic, heterotrophic or parasitic	Autotrophic, heterotrophic or parasitic	Heterotrophic	Autotrophic (photosynthetic)	Heterotrophic
Locomotion	Some have flagella	Some have an undulipodium, some have cilia	None	None	Muscular tissue

Now test yourself

TESTED

2 Explain why fungi were once classified as plants, but have been reclassified in their own kingdom.

Answer on p. 120

New classification systems

REVISED

Similarities in biological molecules

More recent approaches involving sequencing of DNA and proteins have been used to classify organisms. This evidence from **biological molecules** provides a **genetic** and biochemical comparison, which is thought to be more accurate than comparing observable features. Those organisms with similar DNA sequences must be closely related. As changes in DNA sequence are caused by mutation and mutations are random, it is assumed that more mutations means a longer time has elapsed since the species became separate. Therefore, the more mutations or differences in the DNA, the less closely related the species.

As the DNA sequencing is used to produce proteins, it is equally valid for studying amino acid sequences in proteins. The best proteins are those that occur in all living things, such as proteins associated with respiration and protein synthesis. These include cytochrome C and RNA polymerase.

Once the taxonomic groups are constructed on the basis of genetic similarities, they will automatically reveal the evolutionary relationships between organisms (phylogeny).

The three domains of life

Until recent years, the five kingdom hierarchy of classification was universally accepted. However, recent research has revealed that this hierarchy may not be accurate. This research included detailed study of the sequence of bases in the RNA of the ribosomes.

One kingdom, the Prokaryotae, consists of a wide diversity of prokaryotic organisms. However, part of this kingdom is significantly different from the rest of the kingdom. This led to the idea that the kingdom should be divided into two major groups or domains: bacteria and archaea. The bacteria are fundamentally different from all other living things. The archaea are still prokaryotic, but they are more similar to eukaryotes.

The **three domains of life** classification of bacteria, archaea and eukaryota is now widely accepted and the domains are a taxonomic level above the five kingdoms.

The relationship between classification and phylogeny

Classification and phylogeny

REVISED

Natural **classification** systems group living things according to their similarities. The more features that are shared between two organisms, the more closely related they are. **Phylogeny** is the evolutionary history or the evolutionary relationships between organisms and groups of organisms. Therefore, natural classification reveals the phylogeny. Phylogeny can be represented in an evolutionary tree such as Darwin's tree of life.

Now test yourself

TESTED

3 Explain why using DNA sequencing to classify organisms automatically matches their phylogeny.

Answer on p. 120

The evidence for the theory of evolution by natural selection

The contributions of Darwin and Wallace

REVISED

Charles Darwin made four observations in proposing his theory:
● Parents tend to produce more offspring than are able to survive.
● Populations tend to remain a constant size.
● Offspring look similar to their parents.
● No two individuals look identical.

From the above observations, Darwin made the following deductions:
● There must be a struggle to exist.
● Some offspring are better adapted to surviving than others.

From the above deductions, Darwin devised the following theory:
● Parents produce too many offspring.
● There must be competition to survive.
● Not all the offspring will survive.

Exam practice answers and quick quizzes at **www.hoddereducation.co.uk/myrevisionnotes**

- Although all offspring look similar to their parents, some are better adapted to their environment.
- The better-adapted offspring survive.
- Those that survive will go on to reproduce.
- The less well-adapted offspring do not survive.
- The better-adapted individuals pass on their features to the next generation.

Darwin called this theory 'evolution by **natural selection**'.

At the same time that Darwin was piecing his **theory of evolution** together, another scientist was doing the same thing. **Alfred Russel Wallace** collated evidence from parts of southeast Asia.

> **Natural selection** is selection for certain features by natural forces.

Evidence for evolution

REVISED

There are many lines of evidence for the theory of evolution. These include fossil and molecular evidence in the form of the structure of DNA and other molecules, particularly proteins.

Fossil evidence

Darwin used a lot of evidence from **fossils** to back up his theory of natural selection. Fossils take two forms:
- imprints of ancient organisms
- the remains of organisms that have died and become mineralised

Fossils are found in sedimentary rocks. They are formed when an organism leaves an imprint in soft mud or dies and comes to rest in the mud. As the mud hardens to form rock, the imprint or body remains in the rock. Fossils formed relatively recently are near the surface of modern rocks, whereas those formed many millions of years ago are found further below the surface.

Scientists study fossils in minute detail and carefully describe their anatomy and morphology. Similarities between fossils can be used to reveal evolutionary relationships (phylogeny).

Molecular evidence

Certain chemicals, such as **DNA**, proteins and RNA, are universal to all living things. Variation is caused by changes (mutations) in the DNA, which produces changes in proteins. As members of the same species have similar DNA, they also have similar proteins. As evolution occurs and one species becomes two, the DNA accumulates more changes, as does the structure of the proteins it codes for. Therefore, closely related species have similar DNA and proteins, but more distantly related species have DNA and proteins that are more different.

Sequencing the bases in DNA and the amino acids in proteins shows the similarities and differences between species. This reveals their evolutionary relationships in the same way that similarities in anatomy and morphology reveal evolutionary relationships.

The evidence from biochemistry is, perhaps, more reliable and more convincing than that from fossils.

Different types of variation

> **Variation** is the differences that arise between living organisms.

The term **variation** refers to the differences that exist between individuals.

Intraspecific and interspecific variation

REVISED

Intraspecific variation occurs between members of the same species. These differences could be:
- simple observable features such as colour
- biochemical differences such as the precise sequence of amino acids in a protein
- behavioural differences such as the type of food eaten

These differences are usually relatively minor, but they can be more obvious such as the differences between the sexes.

Variation can also occur between members of different species (**interspecific variation**). This depends on how closely related one species is to the other:
- If the species are closely related, such as the lion and the tiger, the differences may not be great.
- If the species are not closely related, the differences will be greater.

> **Exam tip**
>
> Remember that variation is the key to evolution — variation must occur before any characteristic can become beneficial and selected for.

Continuous and discontinuous variation

REVISED

Continuous variation is seen where there are no distinct groups or categories. There is a full range between two extremes. This form of variation is caused by:
- a number of genes interacting together
- the environment

Examples of continuous variation include height and body weight. The continuously variable feature can be quantified and data are usually presented in the form of a histogram.

Discontinuous variation is seen where there are distinct groups or categories and there are no in-between types. This type of variation is usually caused by one gene (or possibly a small number of genes). The discontinuously variable feature cannot be quantified — it is qualitative — and data are usually presented in the form of a bar chart. Examples of discontinuous variation include gender and possession of resistance or immunity.

> **Continuous variation** is variation that shows a complete range with no distinct groups.

> **Discontinuous variation** is variation that produces distinct groups.

> **Typical mistake**
>
> Students tend to plot bar charts with no spaces between the bars, but there should be gaps between them.

> **Revision activity**
>
> Write a list of features that are continuously variable and a separate list of features that are discontinuously variable.

Causes of variation

REVISED

There are two causes of variation: genetic and environmental. Many variable features may be affected by both causes. For example, skin colour in humans is genetically determined. However, exposure to the sun results in the production of extra pigmentation, causing the skin to tan.

Genetic causes

Genetic causes of variation are a result of differences in the sequence of bases in the DNA. They are caused by mutations that arise spontaneously and randomly, and are passed on from one generation to the next. They usually cause discontinuous variation. Examples include:
- number of limbs
- eye colour
- ability to roll the tongue

> **Now test yourself**
>
> 4 Explain how genetic differences cause visible variation between members of the same species.
>
> Answer on p. 120
>
> TESTED

Exam practice answers and quick quizzes at **www.hoddereducation.co.uk/myrevisionnotes**

Environmental causes

Environmental causes of variation are caused by variations in exposure to certain environmental conditions. They are not passed from one generation to the next and cause continuous variation. Examples include:
- skin colour resulting from exposure to sunlight
- body mass

Adaptations of organisms to their environment

All members of a species possess similar **adaptations**, which enable the species to survive and thrive in their environments. These adaptations can be categorised as anatomical, physiological or behavioural.

Adaptations are features that help organisms to survive in their habitat.

Anatomical adaptations

REVISED ☐

Anatomical adaptations are those that are associated with structure:
- Predators have sharp teeth to help kill and chew their prey. They also have a strong jaw joint so that it does not become dislocated by a struggling victim.
- Herbivores have a long and complex digestive system. This allows them to digest plant tissues, which are much more difficult to digest than animal tissues.
- Plants have long, deep roots with many root hairs. This enables them to absorb water and minerals from the soil.
- Some plants such as the black mangrove, which lives in waterlogged soil, grow roots up into the air. This allows the roots to gain oxygen from the air above the anaerobic soil.

Physiological adaptations

REVISED ☐

Physiological adaptations are those that are associated with how the body systems function:
- Fish pass water over their gills in one direction, as opposed to mammals which have a tidal flow of air into their lungs. This is because water is much more dense and difficult to slow down and stop so that the direction of movement can be reversed.
- The kidneys of mammals extract water from the urine before excreting nitrogenous waste. This helps to reduce the need to find and drink water.
- Some plants, called C4 plants, use an unusual way of collecting (fixing) carbon dioxide at night. This means they can keep their stomata closed during the hottest part of the day, reducing loss of water via transpiration, but are still able to photosynthesise.
- Yeasts respire anaerobically when there is no oxygen in their habitat. This means they can produce ATP and continue to grow.

Behavioural adaptations

REVISED ☐

Behavioural adaptations are those that are associated with feeding, nesting and mating:
- Robins usually choose a nest site in a hole in a tree stump or wall a few inches above the ground. This means they are not competing with other bird species.

- In dry conditions, some plants open their stomata to make the leaves wilt. This reduces the surface area exposed to hot sun and reduces the rate of transpiration.

Convergent evolution

REVISED

In some cases, organisms from different taxonomic groups become adapted to the same habitat by adopting **similar anatomical features**. For example, the **marsupial mole** and the **placental mole** look remarkably similar although they are not closely related. They have independently evolved similar traits as a result of having to adapt to similar environments or ecological niches.

> **Typical mistake**
>
> Students often forget to explain how each adaptation helps survival of the species.

> **Exam tip**
>
> In examination questions, you are likely to be given some information about a particular species and its habitat. The question will then ask you to explain how the features described help the organism to survive.

How natural selection can affect the characteristics of a population

The mechanism of natural selection

REVISED

Genetic variation exists between individuals and some factor in the environment applies a **selection pressure**. Some variations are better adapted to survive than others, passing on their alleles to the next generation in greater numbers. Over a number of generations, the proportion of the population possessing these **advantageous characteristics** increases.

> **Now test yourself**
>
> TESTED
>
> 5 Draw a flow diagram to explain how a change in the environment can cause a change in the proportion of individuals possessing a particular characteristic.
>
> **Answer on p. 120**

The implications of evolution for humans

Modern examples of evolution include pesticide resistance in insects and drug resistance in microorganisms.

Pesticide resistance in insects

REVISED

When pesticides are used, they kill all susceptible **insects**. However, some individuals may have a degree of **pesticide resistance**. These few individuals may survive and breed to pass on their resistance to following generations. The pesticide has acted as a selective agent. As successive generations show some variation, it is possible for the insects to become increasingly resistant to higher and higher concentrations of the pesticide. The implications are that insects can damage food crops and, with no effective pesticides, we will be unable to prevent this damage.

Drug resistance in microorganisms

Most bacteria are susceptible to antibiotics, but some **microorganisms** may show **drug resistance**. Just as with insect resistance to pesticides, the bacterial species eventually evolves resistance. The implications are that there are only a certain number of antibiotics and, once bacteria have evolved resistance to them all, we will have no further defences to help us combat disease.

Now test yourself

6 Explain how strains of the bacterium *Clostridium difficile* can become resistant to antibiotics.

Answer on p. 120

Exam practice

1 The leopard *Panthera pardus* is a member of the cat family. Complete the following table to show its full classification. [5]

Kingdom	
	Chordata
Class	Mammalia
	Carnivora
Family	Felidae
Genus	
	pardus

2 (a) (i) State two features of continuous variation. [2]
    (ii) State two features of discontinuous variation. [2]
  (b) List the causes of variation. [2]
  (c) Explain what is meant by the term *selection*. [3]
3 (a) Explain how inappropriate use of antibiotics has given rise to the so-called superbug methicillin-resistant *Staphylococcus aureus* (MRSA). [5]
  (b) Explain how the structure of proteins can be used as evidence for evolution. [4]

Answers and quick quiz 13 online

Exam tip

Questions sometimes take the form of a description of the mechanism of evolution. This is likely to be in the context of artificial selection of farm animals or plants.

Exam tip

Remember that for new strains or new species to arise, variation and selection must occur over a number of generations.

Revision activity

Write a list of all the key terms used in this chapter, then add the meaning of each key term.

Summary

By the end of this chapter you should be able to:
- Describe the classification of species into the taxonomic hierarchy.
- Outline the characteristic features of the five kingdoms.
- Outline the use of the binomial system for nomenclature.
- Describe the evidence used in the classification of organisms.
- Discuss the fossil and biochemical evidence for evolution.
- Discuss variation within and between species.

- Describe the differences between continuous and discontinuous variation.
- Explain both genetic and environmental causes of variation.
- Outline the adaptations of organisms to their environments.
- Outline the roles of variation, adaptation and selection in evolution.
- Discuss the evolution of resistance to pesticides in insects and resistance to drugs in microorganisms.

Now test yourself answers

Chapter 2

1 $I = A \times M$, $A = \dfrac{I}{M}$

2 The resolution of a light microscope is not sufficient to separate objects that are closer than 200 nm. Therefore, an image magnified at greater than 1500× contains no extra detail.

3 A scanning electron microscope image is in three dimensions and a surface view. A transmission electron microscope can see details inside organelles.

4 There are 1000 nanometers (nm) in one micrometer (μm) and 1000 micrometers in one millimetre.

5 The organelles could be cut in a different plane. Organelles, and especially mitochondria, can change shape and be larger in more active cells.

Chapter 3

1 A δ^+ charge is a small charge not equivalent to a whole electron or proton charge. It is created by polarity in the molecule. Certain atoms attract electrons more strongly than others and if electrons are attracted away from one part of the molecule it leaves a δ^+ charge.

2 (a) Specific heat capacity is a measure of how much energy is needed to warm up a substance. The energy is used to make the molecules move about more. Latent heat capacity is a measure of how much energy is needed to convert a substance from one state (e.g. liquid) to another (e.g. gas). The energy is used in breaking the bonds that hold the molecules together.

   (b) Water has a high specific heat capacity, which means that a lot of energy is needed to warm up a body of water. As living things are mostly water, this means that a body remains at a fairly constant temperature. Latent heat is the heat needed to evaporate water or sweat. Evaporating water off plant leaves and sweat off the skin helps to keep living things cool.

3 Amylose is a long chain of glucose. Glucose holds energy. Amylose is large and therefore insoluble, so it does not affect the water potential of the cell. The molecule is highly coiled, so it does not take up much space. Amylose can be broken down easily by hydrolysis to release the glucose units when energy is needed.

4 Cellulose is a large molecule. It has many hydroxyl groups that can form hydrogen bonds and these bind to other cellulose molecules, making fibrils.

5 Lipids are high energy compounds. The fatty acid chains can be broken down and enter respiration to release energy. Lipids are not water soluble, so they do not affect the water potential of the cell.

6 Phospholipids have a phosphate group attached to the 'head' end. This is hydrophilic and will mix with water — it will twist towards water. The fatty acid 'tails' are hydrophobic and will not mix with water — they twist away from water. A group of phospholipid molecules will orientate so that they form a globule with the heads on the outside in contact with the water and the tails on the inside acting as a barrier to the water.

7 Proteins are chains of amino acids joined by peptide bonds. A peptide bond forms between the amino end of one amino acid and the carboxylic acid end of another. Each amino acid has just one amino group and one carboxylic acid group. Therefore, it cannot form an extra bond to create a branch.

8 A polypeptide is a long chain of amino acids joined by peptide bonds. A protein may consist of one polypeptide chain or it may have several polypeptide chains. A protein may also contain a non-protein prosthetic group such as the haem group in haemoglobin.

9 Metabolically active proteins need a specific shape, which is achieved by the three-dimensional tertiary structure. This gives hormones a binding site and enzymes an active site.

10 The three-dimensional shape of globular proteins is usually very specific. If it is altered by a change in temperature, the protein becomes inactive.

11 A solution with low concentration of glucose contains less glucose than one with high concentration. If the Benedict's reagent is in excess, the low concentration of glucose will react with only some of the available Benedict's reagent, leaving some remaining. If there is a lot of glucose present, it will react with more of the Benedict's reagent, leaving less or none remaining.

Chapter 4

1 If any errors were made in the copying of the base sequence, the gene code would be changed. This is a mutation. A mutation may not produce the required protein or the protein may not function properly.

2 Reading single bases would give four codes. Reading pairs would give 16 codes. Reading triplets gives 64 codes. There must be at least 20 codes as there are 20 amino acids.

3 The DNA is a large molecule — too large to leave the nucleus. mRNA is a smaller molecule and single-stranded, so it will fit through the nuclear pores.

4 mRNA is made by matching nucleotides to the exposed bases in the DNA. It is then used as a template to line up the amino acids in the correct sequence. This can be achieved only if the bases in the mRNA are exposed so that similar unpaired and exposed bases in the tRNA can bind.

Chapter 5

1 You can imagine the need for activation energy as a boulder sitting in a hollow at the top of a hill. The boulder will not roll down the hill until it is pushed up and out of the hollow. Once over the lip of the hollow, the boulder can then roll easily down the hill. An enzyme can provide an alternative route for the reaction — removing the lip of the hollow.

2 The shape of the active site depends on the interactions between the R groups in the enzyme molecule. These interact through bonds such as ionic bonds between oppositely charged parts of the molecule and disulfide bonds between sulfur atoms. There are also weaker hydrogen bonds and interactions such as hydrophilic and hydrophobic interactions and forces of attraction and repulsion between charges. As a substrate molecule enters the active site, the physical presence of the molecule can interfere with the weaker interactions in the active site. Also, any charges on the substrate molecule will interact with charges in the active site and this can cause the shape of the active site to change slightly.

3 In any practical procedure involving enzymes, there may be many possible variables. Each variable has the potential to alter the activity of the enzyme and therefore change the rate of reaction. If two or more variables are changed at the same time, both may have an effect on the rate of reaction. It is not possible to distinguish between the effects that each variable has on the rate of reaction unless the experimenter can be certain that only one variable has been changed.

4 Variables such as temperature and pH need to be controlled so that they do not change during the practical test and affect the rate of reaction. A control is a separate test set up to demonstrate that a particular factor is required for the reaction to occur. It is usual to set up a control in which the enzyme, or source of enzyme, is missing to show that the enzyme is essential for the reaction to occur.

5 A non-competitive inhibitor binds to a part of the enzyme molecule away from the active site. The presence of the inhibitor modifies the pattern of bonds and interactions within the enzyme molecule. As a result, the active site changes shape and is no longer complementary to the shape of the substrate. The substrate cannot enter the active site and no enzyme substrate complexes are formed.

Chapter 6

1 Active transport (using energy from ATP) moves molecules or ions across a membrane. If the ions cannot diffuse back as quickly as they are pumped then a concentration gradient will be formed.

2 Compartments can specialise — they can keep all the molecules and enzymes associated with one process together where they are in higher concentration and are not interfered with by other processes.

3 The phospholipid bilayer stops movement of ions and polar molecules. Transport proteins (carrier proteins and channel proteins) have attachment sites for specific molecules.

4 A signal molecule may be specific to a certain target cell. It also has a specific shape. If the receptor has a specific shape that is complementary to the shape of signal molecule, the signal molecule will be able to bind only to that receptor site and will affect only that particular type of target cell.

5 The phospholipid bilayer is impermeable to charged molecules and ions. They cannot diffuse through the bilayer, so they must have an alternative route. This is through proteins that span the membrane, so it is known as facilitated diffusion.

6 Proteins are large molecules and cannot easily diffuse through the phospholipid bilayer. Also, they are too large to fit through protein pores.

7 The cellulose cell wall is strong and can withstand the pressure of water in the cell.

8 A strong salt solution has a low water potential. The cell will have a higher water potential. As water moves down the water potential gradient from inside the cell to outside, the cell loses

water. Water will no longer push against the sides of the cell and it will lose turgidity.

9 The solution from outside the cell will fill the gap between the cell wall and the plasma membrane. The wall is fully permeable so the solution simply moves in through the wall.

Chapter 7

1 Mitosis produces two daughter cells that are genetically identical to the original cell. It must make a copy of the DNA so that there is one full set of genes to enter each daughter cell, otherwise the cells of a multicellular organism would not contain the same DNA and genes.

2 Cell division and growth are energy-requiring processes. Each daughter cell must have enough energy to grow. Each daughter cell must also have sufficient organelles to carry out the requirements of a living cell.

3 The end product of meiosis is four cells that are haploid. The DNA replicates before division. This means there is twice as much DNA in the cell. It must therefore divide twice to reduce the amount of DNA to half the normal amount (the haploid state).

4 Cells divide and then differentiate. Differentiation turns off some of the genes so that they are not expressed. The cell than changes shape and modifies its contents to be able to carry out its function — this is called specialisation.

5 Many cells that are all specialised in the same way form a tissue. They cannot do much on their own as they are so small. However, a large number of cells all performing the same task can achieve much more together. For example, one muscle cell could not move an organism, but many muscle cells together can generate enough force to move a limb.

Chapter 8

1 Size, surface area to volume ratio, level of activity and metabolic rate.

2 An amoeba is small, it has a large surface area to volume ratio and it can absorb all the oxygen it needs through its surface. A large tree has a small surface area to volume ratio, so its surface area is too small to supply the volume. Also, diffusion is too slow to allow oxygen to reach all parts of the body.

3 Concentration on the supply side (needs to be high), concentration on the demand side (needs to be low), thickness of barrier (needs to be thin).

4 To prevent air entering the system or leaving the system through the nose. This would alter the volume of air in the chamber and lead to inaccurate readings.

5 As the subject exhales, carbon dioxide enters the chamber. A high concentration of carbon dioxide alters the breathing rate and may become harmful. Soda lime absorbs the carbon dioxide. This causes the volume in the air chamber to decrease and allows the volume of oxygen used to be calculated.

Chapter 9

1 Double circulation ensures that the oxygenated blood is separated from the deoxygenated blood. This makes transport of oxygen more efficient. Double circulation also allows a higher pressure in the systemic side so that materials can be delivered further and more quickly. The lower pressure in the pulmonary side reduces the chance of damage to the capillaries in the lungs.

2 Artery walls are thick and contain thick layers of muscle and elastic tissue. They also contain collagen. The collagen is strong and prevents the artery wall bursting under high pressure. The inner lining of the artery wall is folded and can unfold as the pressure and volume of blood increase so that the lining does not get damaged or split. As the pressure and volume of blood increase, the wall of the artery stretches to accommodate the extra blood. Once the pressure starts to decrease, the elastic fibres recoil to reduce the diameter of the artery again.

3 Veins have much less smooth muscle and elastic tissue in their walls than arteries. They do not need to stretch and recoil as the pressure is never high. They also have much less collagen. The veins are often flattened and only become round in cross-section when they are full of blood. They also contain valves in their walls which prevent the blood flowing backwards away from the heart.

4 The volume of the ventricle chambers is the same on both sides of the heart. The wall of the left ventricle has much thicker muscle so that it can create a higher pressure to push the blood further around the body.

5 The atrioventricular valves are pushed open and closed by changes in pressure in the blood. When the atrium wall contracts and the pressure in the atrium is higher than in the ventricle, the valves are pushed open as the blood flows from higher to lower pressure. When the ventricle walls contract, the pressure in the ventricles rises above that in the atria and the valves are pushed closed as the blood tries to flow away from the high pressure area.

6 The blood flowing from the atrium to the ventricle moves more slowly than the electrical stimulus. The delay gives time for the blood to flow and fill the ventricle before the ventricle walls are stimulated to contract.

7 The pressure in the ventricle rises above the pressure in the atrium. Blood starts to move back towards the atrium, but it is trapped in the atrioventricular valve and pushes the valve closed.

8 Air contains only about 20% oxygen. As the oxygen in the air is used up, the oxygen tension decreases and the blood in the lungs is less able to extract oxygen from the air. With less saturation of the haemoglobin, less oxygen is transported around the body, so less oxygen is supplied to the brain and respiring tissues.

Chapter 10

1 When water evaporates from the surfaces of the cells in the leaf, it creates a pull or tension on the chain of water molecules in the xylem. This suction effect reduces the pressure in the xylem vessels and they may collapse under the pressure from surrounding tissues. The lignin thickening strengthens the walls to support them against this collapse. It is important that water can continue in an unbroken chain all the way up the xylem. If a xylem vessel becomes blocked or damaged, water can pass through the pits from one vessel to another to maintain the unbroken chain. The pits allow water to pass out of the vessel into the surrounding cells and tissues.

2 Companion cells carry out the active processes that are used to load sucrose into the sieve tube elements.

3 Increasing the temperature increases evaporation, so the water potential in the air spaces of the leaf increases. This increases the water vapour potential gradient between the leaf air spaces and outside the leaf, so water vapour diffuses out of the leaf more quickly. Higher light intensity causes the stomata to open wider to allow more carbon dioxide to enter the leaf for photosynthesis. With wider stomatal openings, more water vapour can diffuse out of the leaf.

4 Transpiration is the loss of water vapour from the aerial parts of the plant. The transpiration stream replaces this loss by transporting water into the roots and up the stem to the leaves.

5 In spring when the leaf buds are opening, they require a supply of sucrose. This is used in respiration to supply the energy needed for growth and also as a building block to make new cells. The leaf is therefore a sink. Once the leaf is formed and can start to photosynthesise, it manufactures sugars that will be transported away to other parts of the plant. The leaf is then a source.

Chapter 11

1 The pathogen may remain on the outside of the vector (on its mouthparts). It will not affect the vector's tissues.

2 Phloem and xylem are transport tissues. The pathogen can be transported around the plant in the mass flow of fluids in these tissues, so blocking them reduces its spread.

3 Blood must not clot in the vessels or it could cause a blockage that may cause a heart attack or stroke. Making the process complex ensures that clots are not formed at inappropriate times.

4 All cells have proteins or glycoproteins on their cell-surface membranes. On the surface of a pathogen, these may have functions such as enabling the cell to bind to other cells or to receptor sites on the host cell membrane. The immune system can recognise these proteins or glycoproteins and use them as antigens to recognise foreign cells.

5 Injecting antibodies provides instant immunity but it is short term — the antibodies do not last long in the body. No memory cells are made, so there is no long-term immunity. If dead cells or antigens are injected, this allows the immune system to become activated and produce antibodies itself along with memory cells. These memory cells are what provide long-term immunity. However, there will be only a limited number of pathogen cells or antigens and the immune system may not be fully activated. If living cells are injected, they will reproduce and increase in number — they mimic a real infection. This activates the immune system more fully and provides complete immunity.

Chapter 12

1 Any bias in selecting sites for samples will make the results less valid. This is because the element of choice can be affected by what the sampler sees. If he or she chooses an area with fewer species, the overall measurement of diversity will be an underestimate. If he or she chooses an area with more species, the overall estimate will be an overestimate. It would also be tempting to select areas that contain rare species, which would affect the decision about how important the habitat is as an environment for rare organisms.

2 Species richness is a measure of how many different species are found in a habitat. If there are a lot of species, the habitat is more diverse and likely to be more stable. Species evenness is a measure of how many individuals there are of each species. If all the species in a habitat are represented by a large

number of individuals or a large population, the population is likely to be stable. It also means that the habitat is more diverse.

3 Monitoring diversity over a number of years can reveal trends and changes in the populations of organisms. Careful monitoring with comparisons to physical factors on the site can help to gain an understanding of the environmental factors that affect the growth and survival of organisms. If the site is likely to be affected by a development, it may be important to establish if there are rare or endangered species in the area in order to ascertain if the site is of importance for conservation.

4 Diverse habitats contain more species and are more stable and likely to survive some level of disruption. When resources are limited and only certain areas can be conserved, it is more important to conserve a habitat containing more species rather than one that contains few species or is unstable and could be damaged despite the attempts to conserve it.

5 Keeping two or more populations of a species allows independent evolution that may enhance genetic variation. Occasional cross-breeding between the populations reduces the chances of harmful genetic combinations arising through inbreeding. If one population is harmed by a disease, the second population ensures that the whole species is not wiped out. Having two populations also benefits scientific research into the species as comparisons can be made between the two populations.

6 Wild animals and plants do not respect international boundaries. If one country spends a lot of effort on conserving a species, the species does not benefit if another country allows the wholesale slaughter of that species. If trade in an endangered species is allowed in one country, this encourages poaching in the natural habitat of another country that is trying to conserve the species.

Chapter 13

1 All members of one species belong to the same genus, which in turn belongs to the same family, order, class etc. As you move up the hierarchy, the taxonomic groups get bigger.

2 Fungi have a spreading body structure like plants and are not able to move around like plants. They also have nuclei and membrane-bound organelles like plants. However, fungi do not possess chloroplasts and are not able to photosynthesise — they are not autotrophic.

3 If the DNA of two species is similar, there have been few mutations to make changes in the DNA. As mutations are spontaneous and random, there will be more mutations and more differences in the DNA between two species that have been evolving separately for a long time, i.e. they are not closely related. If there are few differences in the DNA, the species have not been separate for a long time and they are therefore closely related.

4 Genes code for the structure of proteins and for the sequence of bases in the protein. The protein contributes to the formation of a particular feature. If the base sequence in the DNA changes, the code also changes. This alters the sequence of amino acids in the protein. The new protein may not work or may contribute to the production of a different visible feature.

5 There is a change in the environment → which causes selective pressure → due to natural variation between members of the species → some individuals possess a feature that gives them an advantage → these individuals survive more easily → they reproduce successfully to pass on their genes.

6 A patient infected with *Clostridium difficile* is given antibiotics to kill the bacterium. Due to natural variation, some of the bacteria have some resistance. If the whole course of antibiotics is not completed, some of the resistant bacteria will survive. These reproduce and pass on the gene for resistance to future generations. The next generation also shows variation and some individuals are more resistant than others. Over many generations of bacterial growth and reproduction, successive generations will become increasingly more resistant to the antibiotic.